[英] 戈弗雷·哈代 著
何生 译

一个数学家的辩白（双语版）
A Mathematician's Apology

人民邮电出版社
北京

图书在版编目（CIP）数据

一个数学家的辩白：双语版：汉英对照 /（英）戈弗雷•哈代著；何生译. -- 北京：人民邮电出版社，2020.1（2024.3重印）
（图灵新知）
ISBN 978-7-115-52347-1

Ⅰ. ①一… Ⅱ. ①戈… ②何… Ⅲ. ①数学－普及读物－汉、英 Ⅳ. ①O1-49

中国版本图书馆CIP数据核字(2019)第230282号

内 容 提 要

本书是哈代于1940年写成的心得之作，展现了数学之美、数学的持久性和数学的重要性三大主题。作者从自己的角度谈论了数学中的美学，给众多数学"门外汉"一个机会，洞察数学家的内心。作者还讨论了数学的本质与特点、数学的历史及其社会功能等诸多话题。本书被称为"用优雅的语言对数学真谛进行的完美揭示"，原汁原味地向读者展示了一位真正、纯粹的数学家的数学思想，是不可多得的经典读物。

- ◆ 著　　[英] 戈弗雷•哈代
  译　　　何　生
  责任编辑　傅志红
  责任印制　周昇亮
- ◆ 人民邮电出版社出版发行　北京市丰台区成寿寺路11号
  邮编　100164　电子邮件　315@ptpress.com.cn
  网址　http://www.ptpress.com.cn
  北京天宇星印刷厂印刷
- ◆ 开本：880×1230　1/32
  印张：4.75　　　　　　2020年1月第1版
  字数：133千字　　　　2024年3月北京第11次印刷

定价：59.80元
读者服务热线：(010)84084456-6009　印装质量热线：(010)81055316
反盗版热线：(010)81055315
广告经营许可证：京东市监广登字 20170147 号

献给约翰·洛马斯[1]，本书受他之约而作

Preface

I am indebted for many valuable criticisms to Professor C. D. Broad and Dr C. P. Snow, each of whom read my original manuscript. I have incorporated the substance of nearly all of their suggestions in my text, and have so removed a good many crudities and obscurities.

In one case I have dealt with them differently. My §28 is based on a short article which I contributed to *Eureka* (the journal of the Cambridge Archimedean Society) early in the year, and I found it impossible to remodel what I had written so recently and with so much care. Also, if I had tried to meet such important criticisms seriously, I should have had to expand this section so much as to destroy the whole balance of my essay. I have therefore left it unaltered, but have added a short statement of the chief points made by my critics in a note at the end.

G. H. H

18 July 1940

序

我非常感谢查理·布罗德[2]教授和查尔斯·斯诺[3]博士对本书提出了许多有价值的批评意见,他们都读过我的手稿。他们提出的几乎所有建议的精髓都已被我采纳,我借此改正了书中许多经不起推敲或是令人费解的地方。

在处理那些建议时,我对书中的第 28 节采取了不同的做法,它是基于我年初发表在《顿悟》杂志(剑桥阿基米德协会的期刊)上的一篇短文[4]写成的。我发现自己几乎没有办法去改写这篇文章,它是我精心写就的新作。倘若我认真回复那些重要批评,就会把那一节无限扩写,乃至于会破坏随笔的总体平衡。因此,我保留了第 28 节的原貌,只在注记中补充了一小段文字,探讨他们提出的主要观点。

戈弗雷·哈代

1940 年 7 月 18 日

1

It is a melancholy experience for a professional mathematician to find himself writing about mathematics. The function of a mathematician is to do something, to prove new theorems, to add to mathematics, and not to talk about what he or other mathematicians have done. Statesmen despise publicists, painters despise art-critics, and physiologists, physicists, or mathematicians have usually similar feelings; there is no scorn more profound, or on the whole more justifiable, than that of the men who make for the men who explain. Exposition, criticism, appreciation, is work for second-rate minds.

I can remember arguing this point once in one of the few serious conversations that I ever had with Housman. Housman, in his Leslie Stephen lecture *The Name and Nature of Poetry*, had denied very emphatically that he was a 'critic'; but he had denied it in what seemed to me a singularly perverse way, and had expressed an admiration for literary criticism which startled and scandalized me.

He had begun with a quotation from his inaugural lecture, delivered twenty-two years before—

> Whether the faculty of literary criticism is the best gift that Heaven has in its treasures, I cannot say; but Heaven seems to think so, for assuredly it is the gift most charily bestowed. Orators and poets..., if rare in comparison with blackberries, are commoner than returns of Halley's comet: literary critics are less common... .

1

让职业数学家去写一本关于数学的书,他一定会很发愁。数学家的工作应该是去证明新定理、发现新数学,不该谈论自己或其他数学家做了什么。政治家看不起时事评论家,画家轻视艺术评论家,生理学家、物理学家、数学家们通常也有类似的感觉:这是实干家对评论家的藐视,没有比这种藐视更深刻,或总体来说更无可非议的了。解说、评论、品鉴,都是二等人才从事的工作。

我记得,在和豪斯曼[5]为数不多的几次认真谈话里,就有一次对这个话题展开过辩论。豪斯曼在他的莱斯利·斯蒂芬[6]讲座《诗歌的名与实》上,坚决不承认自己是一个"批评家"。在我看来,他表达的方式很荒谬,其对文学批评表示的赞赏,也让我非常震惊。

他以 22 年前就职演说中的一段话作为开头:

> 我不能说,文学批评能力是否是上天赐予我们的最好礼物。但上天似乎是这样认为的,毫无疑问,它是一份最谨慎的馈赠。演说家和诗人……虽然不像随处可见的黑莓,但他们可比哈雷彗星的回归来得常见,而文学批评家则更稀缺……

And he had continued—

In these twenty-two years I have improved in some respects and deteriorated in others, but I have not so much improved as to become a literary critic, nor so much deteriorated as to fancy that I have become one.

It had seemed to me deplorable that a great scholar and a fine poet should write like this, and, finding myself next to him in Hall a few weeks later, I plunged in and said so. Did he really mean what he had said to be taken very seriously? Would the life of the best of critics really have seemed to him comparable with that of a scholar and a poet? We argued these questions all through dinner, and I think that finally he agreed with me. I must not seem to claim a dialectical triumph over a man who can no longer contradict me; but 'Perhaps not entirely' was, in the end, his reply to the first question, and 'Probably no' to the second.

There may have been some doubt about Housman's feelings, and I do not wish to claim him as on my side; but there is no doubt at all about the feelings of men of science, and I share them fully. If then I find myself writing, not mathematics but 'about' mathematics, it is a confession of weakness, for which I may rightly be scorned or pitied by younger and more vigorous mathematicians. I write about mathematics because, like any other mathematician who has passed sixty, I have no longer the freshness of mind, the energy, or the patience to carry on effectively with my proper job.

他继续说道：

> 在这 22 年里，我在某些方面有所进步，不过在另一些方面退步了。但我还没有进步到足以成为一名文学批评家；同样，我也没有退步到幻想自己已经是一名文学批评家。

一位伟大的学者和优秀的诗人竟然这样认为，在我看来是很可悲的。几个星期后，当我在大厅里发现旁边坐的是豪斯曼时，便单刀直入地和他聊起了以下话题：他的话当真吗？在他看来，最好的评论家真的能与学者和诗人相提并论吗？整个晚餐，我们都在辩论这些问题，我想他最终同意了我的观点。对一个再也无法反驳我的人[7]，我似乎并不能宣布这次辩论取得了胜利。不过，最终他对第一个问题的回答是"也许不能完全当真"，对第二个问题的回答是"或许不能相提并论"。

人们对豪斯曼的感受可能还有些不解，我也不指望他和我的想法是一致的。但科学家的感受是毋庸置疑的，我和他们有完全相同的体会。当我发现自己的创作只不过与数学"有关"，而并不是数学本身时，那就是在承认自己不行了，我很可能会因此而遭受更年轻、更有活力的数学家的轻视或怜悯。就像其他任何一位年逾花甲的数学家一样，我围绕着数学写作，是因为头脑已经老化，不再有足够的精力和耐心去有效地从事数学本职工作了。

2

I propose to put forward an apology for mathematics; and I may be told that it needs none, since there are now few studies more generally recognized, for good reasons or bad, as profitable and praiseworthy. This may be true; indeed it is probable, since the sensational triumphs of Einstein, that stellar astronomy and atomic physics are the only sciences which stand higher in popular estimation. A mathematician need not now consider himself on the defensive. He does not have to meet the sort of opposition described by Bradley in the admirable defence of metaphysics which forms the introduction to *Appearance and Reality*.

A metaphysician, says Bradley, will be told that 'metaphysical knowledge is wholly impossible', or that 'even if possible to a certain degree, it is practically no knowledge worth the name'. 'The same problems,' he will hear, 'the same disputes, the same sheer failure. Why not abandon it and come out? Is there nothing else more worth your labour?' There is no one so stupid as to use this sort of language about mathematics. The mass of mathematical truth is obvious and imposing; its practical applications, the bridges and steam-engines and dynamos, obtrude themselves on the dullest imagination. The public does not need to be convinced that there is something in mathematics.

All this is in its way very comforting to mathematicians, but it is

2

我打算替数学做一次辩白。也许有人会和我说,数学根本不需要这些,因为当下很少有研究工作能像数学一样,无论出于什么原因,都能被公认为是有益的,并且也值得称道。这也许是真的。事实上,由于爱因斯坦[8]激动人心的成果,在大众眼里,可能只有恒星天文学和原子物理学的地位会比数学高。数学家不必认为自己正处于守势,也不需要面对像布拉德利[9]在维护形而上学时所做的辩白里描述的那种敌意,那份令人钦佩的辩白就是《现象与实在》的引言。

据布拉德利说,人们会对形而上学家说,"形而上学的知识是根本不存在的",或是"即便在某种情况下它们是存在的,但实际上它们还是没有什么名副其实的内容"。还有人会说:"同样的问题,同样的争论,同样的彻底溃败。为什么不另起炉灶呢?难道没有别的事情值得去做了吗?"没有人会愚蠢到对数学说这种话。大量数学真理的权威性是明摆着的。它的实际应用随处可见,桥梁、蒸汽机和发电机都是例子。不用唠叨,人们就知道数学很有用。

在某种程度上,所有这些都能让数学家感到欣慰,但真正的数

hardly possible for a genuine mathematician to be content with it. Any genuine mathematician must feel that it is not on these crude achievements that the real case for mathematics rests, that the popular reputation of mathematics is based largely on ignorance and confusion, and that there is room for a more rational defence. At any rate, I am disposed to try to make one. It should be a simpler task than Bradley's difficult apology.

I shall ask, then, why is it really worth while to make a serious study of mathematics? What is the proper justification of a mathematician's life? And my answers will be, for the most part, such as are to be expected from a mathematician: I think that it is worth while, that there is ample justification. But I should say at once that my defence of mathematics will be a defence of myself, and that my apology is bound to be to some extent egotistical. I should not think it worth while to apologize for my subject if I regarded myself as one of its failures.

Some egotism of this sort is inevitable, and I do not feel that it really needs justification. Good work is not done by 'humble' men. It is one of the first duties of a professor, for example, in any subject, to exaggerate a little both the importance of his subject and his own importance in it. A man who is always asking 'Is what I do worth while?' and 'Am I the right person to do it?' will always be ineffective himself and a discouragement to others. He must shut his eyes a little and think a little more of his subject and himself than they deserve. This is not too difficult: it is harder not to make his subject and himself ridiculous by shutting his eyes too tightly.

学家几乎不可能会对此感到满意。任何一位真正的数学家一定会认为，数学的口碑所仰仗的并不是这些朴素的实际应用成果，它在很大程度上是出于人们的无知和不解，所以还有更合理的辩词。无论如何，我打算试一试。相较于布拉德利艰难地为形而上学辩白，这应该会简单些。

那么我得问，为什么认真研究数学的确是值得的呢？数学家存在的意义又是什么呢？在很大程度上，我的答案就是数学家的答案：我认为数学研究是值得的，数学家的存在也是有充分理由的。但同时我还要说明，我为数学的辩白也是在为自己说话，这份辩白在某种程度上必然会很本位。如果我认为自己在数学上很失败，那就不会认为有必要为它辩白。

这种本位主义是不可避免的，我不认为真的需要为此辩解。优秀的成果不是由那些"谦虚"的人做出来的。无论什么学科，教授的首要职责之一，便是把他教的课程以及自己在其中的重要性稍作夸大。一个人若总是问自己"我做的事值得吗？""我是研究这个的合适人选吗？"，那就永远做不好自己，也会让别人情绪低落。他必须不要太在意，稍微拔高一下学科和自身。这点并不难做到，不盲目把它们吹嘘得荒唐可笑才是更难的。

3

A man who sets out to justify his existence and his activities has to distinguish two different questions. The first is whether the work which he does is worth doing; and the second is why he does it, whatever its value may be. The first question is often very difficult, and the answer very discouraging, but most people will find the second easy enough even then. Their answers, if they are honest, will usually take one or other of two forms; and the second form is merely a humbler variation of the first, which is the only answer which we need consider seriously.

(1) 'I do what I do because it is the one and only thing that I can do at all well. I am a lawyer, or a stockbroker, or a professional cricketer, because I have some real talent for that particular job. I am a lawyer because I have a fluent tongue, and am interested in legal subtleties; I am a stockbroker because my judgement of the markets is quick and sound; I am a professional cricketer because I can bat unusually well. I agree that it might be better to be a poet or a mathematician, but unfortunately I have no talent for such pursuits.'

I am not suggesting that this is a defence which can be made by most people, since most people can do nothing at all well. But it is impregnable when it can be made without absurdity, as it can by a substantial minority: perhaps five or even ten per cent of men can do something rather well. It

3

一个人想要证明自己的存在和行为是有意义的,就必须辨别两个不同的问题。第一个问题是,他的工作是否值得去做;第二个则是,无论其价值如何,他为什么要去做。前者通常很难回答,答案也常令人十分沮丧。然而,多数人会觉得回答后一个问题很容易。如果这些人是诚实的,那么答案通常会符合两种形式之一,由于第二种只是比第一种更谦卑,于是第一种形式便成了我们唯一需要认真讨论的答案。

(1)"我做我所做的,因为这是唯一一件我能做好的事。我做律师,或股票经纪人,或职业板球运动员,是因为我真的有天赋来胜任这份工作。我是一名律师,因为我口齿伶俐,并且有志于把控法律的微妙之处;我是一名股票经纪人,因为我对行情的判断既快又准;我是一名职业板球运动员,因为我的击球技术出类拔萃。我同意,成为诗人或数学家也许会更好,但不幸的是我并没有从事这些职业的天赋。"

我并不是说多数人能这样替自己辩解,其实多数人什么都做不好。也许只有5%、最多也就10%的人可以在他的行当里干得相当

is a tiny minority who can do anything *really* well, and the number of men who can do two things well is negligible. If a man has any genuine talent, he should be ready to make almost any sacrifice in order to cultivate it to the full.

This view was endorsed by Dr Johnson—

> When I told him that I had been to see [his namesake] Johnson ride upon three horses, he said 'Such a man, sir, should be encouraged, for his performances show the extent of the human powers…' —

and similarly he would have applauded mountain climbers, channel swimmers, and blindfold chess-players. For my own part, I am entirely in sympathy with all such attempts at remarkable achievement. I feel some sympathy even with conjurors and ventriloquists; and when Alekhine and Bradman set out to beat records, I am quite bitterly disappointed if they fail. And here both Dr Johnson and I find ourselves in agreement with the public. As W. J. Turner has said so truly, it is only the 'highbrows' (in the unpleasant sense) who do not admire the 'real swells'.

We have of course to take account of the differences in value between different activities. I would rather be a novelist or a painter than a statesman of similar rank; and there are many roads to fame which most of us would reject as actively pernicious. Yet it is seldom that such differences of value will turn the scale in a man's choice of a career, which will almost always be dictated by the limitations of his natural abilities. Poetry is more valuable than cricket, but Bradman would be a fool if he sacrificed his cricket in order to write second-rate minor poetry (and I suppose that it is unlikely that

出色。倘若是这一小部分人如此辩解,那么他们的说法一点儿也不荒谬,它是无懈可击的。能真正做好一件事的人非常少,能做好两件事的人更是寥寥无几。如果某人真有天赋,那么为了把这份天赋发挥到极致,他应该做好牺牲一切的准备。

约翰逊博士[10]赞同这个观点:

> 当我告诉他,我曾见过一个和他一样也叫约翰逊的人骑着三匹马时,他说:"先生,这样的人应该受到鼓励,因为他的表演展示了人类力量的极限……"[11]

同样,他也会赞美登山者、横渡海峡的泳将,以及盲棋手。就我个人而言,我完全支持这种为了取得杰出成就而做出的全部努力。即使是魔术师和口技演员,我也能表示理解。在阿廖欣[12]和布拉德曼[13]即将打破纪录的那一刻,倘若他们失败了,我会感到非常失望。就这点而言,约翰逊博士和我都觉得,我们和公众的观点是一致的。正如沃尔特·特纳[14]所说,只有那些"趣味高雅的人"(带有贬义)才不欣赏"真正的名家"。

当然,我们也必须考虑不同活动之间的价值差异。我宁愿当小说家或画家,也不愿做相同级别的政治家。还有许多出名的办法,多数人会因其有问题而加以拒绝。然而,这种价值差异几乎不会改变一个人对职业的选择,择业几乎总是由个人天赋的局限性决定的。诗歌比板球更有价值,但如果布拉德曼为了写二流小诗(我料想他

he could do better). If the cricket were a little less supreme, and the poetry better, then the choice might be more difficult: I do not know whether I would rather have been Victor Trumper or Rupert Brooke. It is fortunate that such dilemmas occur so seldom.

I may add that they are particularly unlikely to present themselves to a mathematician. It is usual to exaggerate rather grossly the differences between the mental processes of mathematicians and other people, but it is undeniable that a gift for mathematics is one of the most specialized talents, and that mathematicians as a class are not particularly distinguished for general ability or versatility. If a man is in any sense a real mathematician, then it is a hundred to one that his mathematics will be far better than anything else he can do, and that he would be silly if he surrendered any decent opportunity of exercising his one talent in order to do undistinguished work in other fields. Such a sacrifice could be justified only by economic necessity or age.

不太可能写出更好的作品）而放弃板球，那他就是个傻瓜。倘若他的板球技巧不那么高超，而诗歌还写得稍好一些，那么可能会更难以抉择：我不知道自己会更愿意成为维克托·特兰佩[15]还是鲁珀特·布鲁克[16]。幸运的是，这样的困境几乎没有出现。

我还可以补充一点，这些人绝不可能想要当数学家。尽管数学家与其他人在思维过程上的差异常常被过分夸大，但不可否认的是，数学才能是一种最专业的天赋，而数学家是这样一类人，他们的常规能力或通才能力并不特别突出。如果一个人不管以什么标准衡量，都能算得上是真正的数学家，那么几乎可以肯定，相较他能做的其他工作，从事数学会好得多。倘若他为了能在其他领域有一份普普通通的工作，而放弃可以发挥自己数学才能的良机，那他就是愚蠢的。只有出于经济或年龄的考虑，这种牺牲才说得过去。

4

I had better say something here about this question of age, since it is particularly important for mathematicians. No mathematician should ever allow himself to forget that mathematics, more than any other art or science, is a young man's game. To take a simple illustration at a comparatively humble level, the average age of election to the Royal Society is lowest in mathematics.

We can naturally find much more striking illustrations. We may consider, for example, the career of a man who was certainly one of the world's three greatest mathematicians. Newton gave up mathematics at fifty, and had lost his enthusiasm long before; he had recognized no doubt by the time that he was forty that his great creative days were over. His greatest ideas of all, fluxions and the law of gravitation, came to him about 1666, when he was twenty-four—'in those days I was in the prime of my age for invention, and minded mathematics and philosophy more than at any time since'. He made big discoveries until he was nearly forty (the 'elliptic orbit' at thirty-seven), but after that he did little but polish and perfect.

Galois died at twenty-one, Abel at twenty-seven, Ramanujan at thirty-three, Riemann at forty. There have been men who have done great

4

关于年龄问题,我最好补充几句,因为它对数学家特别重要。任何一位数学家都不应该让自己忘记,比起任何其他艺术或科学,数学更是年轻人的游戏。举一个相对简单的例子,在英国皇家学会的入选者中,数学家的平均年龄是最小的。

我们还可以很轻松地找到更多引人注目的例证。比如,我们可以看看下面这个人的职业生涯,他无疑是世界上最伟大的三位数学家之一。牛顿[17]在50岁时放弃了数学研究,他在很久以前就失去了对数学的热情;毫无疑问,他在40岁时就意识到他那最富有创造力的数学生涯已经结束。牛顿最伟大的思想——流数术和万有引力定律——是在1666年左右产生的,那时他才24岁。"在那些日子里,我正处于发明创造的黄金时期,我比任何时候都更专注于数学和哲学。"他不断地取得重大发现,一直到将近40岁(他在37岁时算出了"椭圆轨道"),但在此之后,他除了修正和完善之前的成果,几乎再也没有做出什么新的东西了。

伽罗瓦[18]21岁就死了,阿贝尔[19]27岁,拉马努金[20]33岁,黎曼[21]也只活到40岁。也有人在上了年纪之后做出过了不起的成就,

work a good deal later; Gauss's great memoir on differential geometry was published when he was fifty (though he had had the fundamental ideas ten years before). I do not know an instance of a major mathematical advance initiated by a man past fifty. If a man of mature age loses interest in and abandons mathematics, the loss is not likely to be very serious either for mathematics or for himself.

On the other hand the gain is no more likely to be substantial; the later records of mathematicians who have left mathematics are not particularly encouraging. Newton made a quite competent Master of the Mint (when he was not quarrelling with anybody). Painlevé was a not very successful Premier of France. Laplace's political career was highly discreditable, but he is hardly a fair instance, since he was dishonest rather than incompetent, and never really 'gave up' mathematics. It is very hard to find an instance of a first-rate mathematician who has abandoned mathematics and attained first-rate distinction in any other field.* There may have been young men who would have been first-rate mathematicians if they had stuck to mathematics, but I have never heard of a really plausible example. And all this is fully borne out by my own very limited experience. Every young mathematician of real talent whom I have known has been faithful to mathematics, and not from lack of ambition but from abundance of it; they have all recognized that there, if anywhere, lay the road to a life of any distinction.

---

* Pascal seems the best.

高斯[22]关于微分几何的著名论文是在他50岁时发表的（尽管10年前他就有这方面的基本思想）。据我所知，在数学上没有一项重大的进步是由超过50岁的人提出的。如果一把年纪的人丧失了对数学的兴趣并将它抛弃，由此造成的损失对数学和他个人而言都不会很严重。

另外，数学家们在离开数学领域之后的状况也并不那么振奋人心，他们也都没什么实质性的建树。牛顿（在不和别人争吵的时候）是一个相当能干的铸币厂厂长。班勒卫[23]是一位不太成功的法国总理。拉普拉斯[24]的政治生涯极不光彩，但他几乎不能算是一个合适的例子，因为他不是无能，而是不诚实，而且他从来没有真正"放弃"过数学。很难找到第一流的数学家在放弃数学之后，在其他领域取得卓越成就的例子*[25]。也许有一些年轻人，倘若他们专攻数学，就会成为一流的数学家，但我从未听说过一个确实可信的例子。我自己有限的经历反复证明了这一切。我所认识的每一位真正才华横溢的年轻数学家都对数学忠心耿耿，他们志存高远，充满雄心壮志。他们都意识到，如果存在一条可以通往卓越人生的道路，那这条道路就是数学。

---

* 似乎帕斯卡已经是做得最好的了。

5

There is also what I called the 'humbler variation' of the standard apology; but I may dismiss this in a very few words.

(2) 'There is *nothing* that I can do particularly well. I do what I do because it came my way. I really never had a chance of doing anything else.' And this apology too I accept as conclusive. It is quite true that most people can do nothing well. If so, it matters very little what career they choose, and there is really nothing more to say about it. It is a conclusive reply, but hardly one likely to be made by a man with any pride; and I may assume that none of us would be content with it.

5

还有一种答案,我把它称为第一种"更谦卑的变体",我只用几句话一带而过。

(2)"没有什么事我能做得特别出色。我做我现在所做的,因为它刚好就在我面前。我的确从来没有机会去从事别的工作。"我同样认为这个辩解是无法反驳的。的确,大多数人什么也做不好。如果是这样,那么他们选择什么职业是无关紧要的,也确实没什么好说的。这是一个不容置疑的回答,但有自尊心的人不太可能说得出口,我想我们都不会满足于此。

6

It is time to begin thinking about the first question which I put in §3, and which is so much more difficult than the second. Is mathematics, what I and other mathematicians mean by mathematics, worth doing; and if so, why?

I have been looking again at the first pages of the inaugural lecture which I gave at Oxford in 1920, where there is an outline of an apology for mathematics. It is very inadequate (less than a couple of pages), and it is written in a style (a first essay, I suppose, in what I then imagined to be the 'Oxford manner') of which I am not now particularly proud; but I still feel that, however much development it may need, it contains the essentials of the matter. I will resume what I said then, as a preface to a fuller discussion.

(1) I began by laying stress on the *harmlessness* of mathematics—'the study of mathematics is, if an unprofitable, a perfectly harmless and innocent occupation'. I shall stick to that, but obviously it will need a good deal of expansion and explanation.

Is mathematics 'unprofitable'? In some ways, plainly, it is not; for example, it gives great pleasure to quite a large number of people. I was thinking of 'profit', however, in a narrower sense. Is mathematics 'useful', *directly* useful, as other sciences such as chemistry and physiology are? This is not an altogether easy or uncontroversial question, and I shall ultimately

6

现在,是时候开始思考我在第 3 节提出的第一个问题了,它比第二个问题更难。我和其他数学家所说的"数学",真值得去研究吗?如果它是值得的,原因又是什么呢?

在我 1920 年的牛津大学就职演说讲稿的开头几页,里有一份为数学辩白的大纲,我又重温了一遍。辩白的大纲很不完善(只有两页),现在看来,行文风格也不值得夸耀(我想,它是我那时以为颇有"牛津味"的第一篇文章)。但我仍然认为,不管这份大纲需要什么样的改进,它已经囊括了问题的本质。我把当时所说的摘录如下,作为接下来更加全面地讨论这个问题的开场白。

(1) 我在一开始强调了数学的**无害性**——"数学研究即便是无益的,至少也是完全清白无害的工作。"我坚持这一观点,但显然还需要做进一步的展开和解释。

数学**真**的是"无益"的吗?在某些方面,显然不是这样的。例如,它给相当多的人带来了极大的乐趣。不过,我在这里考虑的是狭义上的"益处"。数学是否"有用",它像化学和生理学等其他科学一样**直接**有用吗?这并不是一个非常简单或毫无争议的问题,我

say No, though some mathematicians, and most outsiders, would no doubt say Yes. And is mathematics 'harmless'? Again the answer is not obvious, and the question is one which I should have in some ways preferred to avoid, since it raises the whole problem of the effect of science on war. Is mathematics harmless, in the sense in which, for example, chemistry plainly is not? I shall have to come back to both these questions later.

(2) I went on to say that 'the scale of the universe is large and, if we are wasting our time, the waste of the lives of a few university dons is no such overwhelming catastrophe': and here I may seem to be adopting, or affecting, the pose of exaggerated humility which I repudiated a moment ago. I am sure that that was not what was really in my mind; I was trying to say in a sentence what I have said at much greater length in §3. I was assuming that we dons really had our little talents, and that we could hardly be wrong if we did our best to cultivate them fully.

(3) Finally (in what seem to me now some rather painfully rhetorical sentences) I emphasized the permanence of mathematical achievement—

> What we do may be small, but it has a certain character of permanence; and to have produced anything of the slightest permanent interest, whether it be a copy of verses or a geometrical theorem, is to have done something utterly beyond the powers of the vast majority of men.

And—

> In these days of conflict between ancient and modern studies, there must surely be something to be said for a study which did not begin with Pythagoras, and will not end with Einstein, but is the oldest and the youngest of all.

最终的结论是"没有用",尽管有一些数学家和外行们会斩钉截铁地说它"有用"。那么,数学"无害"吗?同样,这个问题的答案也不是显而易见的,我本该以某种形式回避这个问题,因为它会引出一个大问题,即科学对战争造成的影响。例如,化学显然是有害的,从这种意义上说,数学是无害的吗?这两个问题我后面再谈。

(2) 我接着提到:"宇宙的尺度是巨大的,如果我们是在浪费时间,那么浪费几个大学教师的生命也不算是什么大灾难。"在这里,我似乎采用了一种近乎夸张的谦卑姿态,也就是我刚刚批驳过的那种装腔作势。我明白这并不是自己心里真正想的:我想用一句话来概括我在第 3 节里所谈的冗长内容。我想说的是,我们这些大学教师真的是有些才能的,倘若尽力培养,肯定不会有错。

(3) 最后(现在看来,这些修辞过于夸张),我强调了数学成就的持久性:

> 我们所做的工作也许微不足道,但它具有某种永恒的特性;而产生任何永恒的意义,不管是一首诗还是一个几何定理,哪怕它是最渺小的,都是做到了一件绝大多数人完全无法实现的事情。

此外:

> 在古代成果和现代研究相互冲突的今天,对于一个最古老,同时也是最年轻的研究领域而言——它并非始于毕

All this is 'rhetoric'; but the substance of it seems to me still to ring true, and I can expand it at once without prejudging any of the other questions which I am leaving open.

达哥拉斯[26]，也不会以爱因斯坦作为终点——肯定是有一些可说道的。

所有这些都过于"夸张"。但在我看来，它的本质依然是正确的，我可以立即就此展开，而不影响任何其他有待讨论的问题。

7

I shall assume that I am writing for readers who are full, or have in the past been full, of a proper spirit of ambition. A man's first duty, a young man's at any rate, is to be ambitious. Ambition is a noble passion which may legitimately take many forms; there was *something* noble in the ambition of Attila or Napoleon: but the noblest ambition is that of leaving behind one something of permanent value—

> Here, on the level sand,
> Between the sea and land,
> What shall I build or write
> Against the fall of night?
>
> Tell me of runes to grave
> That hold the bursting wave,
> Or bastions to design
> For longer date than mine.

Ambition has been the driving force behind nearly all the best work of the world. In particular, practically all substantial contributions to human happiness have been made by ambitious men. To take two famous

7

　　我想，我的这篇文章是为那些满怀壮志或曾经充满着雄心的人写的。一个人（至少是一个年轻人）的首要责任，就是要有雄心。雄心是一种高尚的激情，它可以采取多种合理的形式。在阿提拉[27]和拿破仑[28]的雄心里，有着某种崇高的东西。但最崇高的雄心，就是留下一些永恒的价值——

> 这平坦的沙滩，
> 在海陆间漫延，
> 我该营造或写下什么，
> 赶在夜幕降临之前。
>
> 告诉我碑上刻些什么文字，
> 惊涛拍岸也无法将它侵蚀，
> 或是该设计怎样的堡垒，
> 即使我已不在世，它仍坚不可摧。

　　世界上所有优秀成果背后的驱动力，几乎都是雄心壮志。特别是几乎所有对人类幸福做出重大贡献的人，也都是胸怀大志的。李

examples, were not Lister and Pasteur ambitions? Or, on a humbler level, King Gillette and William Willett; and who in recent times have contributed more to human comfort than they?

Physiology provides particularly good examples, just because it is so obviously a 'beneficial' study. We must guard against a fallacy common among apologists of science. The fallacy of supposing that the men whose work most benefits humanity are thinking much of that while they do it, that physiologists, for example, have particularly noble souls. A physiologist may indeed be glad to remember that his work will benefit mankind, but the motives which provide the force and the inspiration for it are indistinguishable from those of a classical scholar or a mathematician.

There are many highly respectable motives which may lead men to prosecute research, but three which are much more important than the rest. The first (without which the rest must come to nothing) is intellectual curiosity, desire to know the truth. Then, professional pride, anxiety to be satisfied with one's performance, the shame that overcomes any self-respecting craftsman when his work is unworthy of his talent. Finally, ambition, desire for reputation, and the position, even the power or the money, which it brings. It may be fine to feel, when you have done your work, that you have added to the happiness or alleviated the sufferings of others, but that will not be why you did it. So if a mathematician, or a chemist, or even a physiologist, were to tell me that the driving force in his work had been the desire to benefit humanity, then I should not believe him (nor should I think the better of him if I did). His dominant motives have been those which I have stated, and in which, surely, there is nothing of which any decent man need be ashamed.

斯特[29]和巴斯德[30]不就是两个比较有名的例子吗？或者层次稍低一些的金·吉列[31]和威廉·威利特[32]，在对人类生活舒适度方面的贡献，近年来又有谁能比得过他们呢？

在生理学方面，有一些特别好的事例，因为它显然是一项"有益"的研究。我们必须警惕在为科学做辩白时出现的一些常见谬论。我料想，那些从事对人类最有益工作的人在工作的时候也会想到这些问题，比方说认为生理学家们拥有特别高尚的灵魂。一个生理学家也许很乐意牢记，他的工作会造福人类，但为其提供动力和灵感的动机，和古典学者或数学家的没什么两样。

有许多令人肃然起敬的动机可以促使人们从事研究工作，但其中有三个比其他的重要得多。首先是求知欲，即对真理的渴求（没有它，其他便无从谈起）。然后是职业自尊心，它是对自己的表现是否能令人满意而生的焦虑，是一个有自尊的手艺人在没能发挥出自己的才能时所面对的羞耻感。最后是雄心，它是对名誉的渴望，是对地位的渴望，甚至是对权力和金钱的渴望。当你完成自己的工作，觉得它为别人增添了快乐或减轻了痛苦时，你可能会感觉很棒，但那并不是你从事那份工作的原因。因此，如果有人告诉我，他工作的动力是为了造福人类，那么不管他是数学家还是化学家，哪怕是生理学家，我也不会相信他（即便相信，我也不会把他想得更好）。他的主要动机一定是我刚刚说的那些，当然，任何体面的人也并不需要为此而感到不好意思。

8

If intellectual curiosity, professional pride, and ambition are the dominant incentives to research, then assuredly no one has a fairer chance of gratifying them than a mathematician. His subject is the most curious of all—there is none in which truth plays such odd pranks. It has the most elaborate and the most fascinating technique, and gives unrivalled openings for the display of sheer professional skill. Finally, as history proves abundantly, mathematical achievement, whatever its intrinsic worth, is the most enduring of all.

We can see this even in semi-historic civilizations. The Babylonian and Assyrian civilizations have perished; Hammurabi, Sargon, and Nebuchadnezzar are empty names; yet Babylonian mathematics is still interesting, and the Babylonian scale of 60 is still used in astronomy. But of course the crucial case is that of the Greeks.

The Greeks were the first mathematicians who are still 'real' to us to-day. Oriental mathematics may be an interesting curiosity, but Greek mathematics is the real thing. The Greeks first spoke a language which modern mathematicians can understand; as Littlewood said to me once, they are not clever schoolboys or 'scholarship candidates', but 'Fellows of

8

如果求知欲、职业自尊心和雄心是从事研究的主要动机，那么毫无疑问，数学家是最有机会具备这些动机的人群了。他们所研究的是所有学科里最令人感到好奇的——没有一门学科所涉及的真理会搞出如此奇怪的把戏。数学学科拥有最精巧和最迷人的技巧，并且为人们提供了绝佳的机会来展示他们纯粹的专业技能。最后，历史还充分证明，不管数学成就的内在价值如何，它都是所有成就中最持久的。

这点甚至在半古文明里也有迹可循。古巴比伦文明和亚述文明早已灰飞烟灭，汉谟拉比[33]、萨尔贡[34]、尼布甲尼撒[35]也都空有其名。然而，古巴比伦的数学依然有其重要性，它的60进制仍在天文学中使用。不过，古希腊人的例子影响更大。

时至今日，能被我们认定为最早的"真正的"数学家，就是那些古希腊人。东方数学可能源于有趣的好奇心，但古希腊数学则是货真价实的数学。古希腊人率先使用了一种现代数学家们能够理解的语言。正如李特尔伍德[36]曾经对我说过的，他们不是机灵的小学生，也不是"奖学金申请人"，而是"某个学院的研究员"。所以古

another college'. So Greek mathematics is 'permanent', more permanent even than Greek literature. Archimedes will be remembered when Aeschylus is forgotten, because languages die and mathematical ideas do not. 'Immortality' may be a silly word, but probably a mathematician has the best chance of whatever it may mean.

Nor need he fear very seriously that the future will be unjust to him. Immortality is often ridiculous or cruel: few of us would have chosen to be Og or Ananias or Gallio. Even in mathematics, history sometimes plays strange tricks; Rolle figures in the text-books of elementary calculus as if he had been a mathematician like Newton; Farey is immortal because he failed to understand a theorem which Haros had proved perfectly fourteen years before; the names of five worthy Norwegians still stand in Abel's *Life*, just for one act of conscientious imbecility, dutifully performed at the expense of their country's greatest man. But on the whole the history of science is fair, and this is particularly true in mathematics. No other subject has such clear-cut or unanimously accepted standards, and the men who are remembered are almost always the men who merit it. Mathematical fame, if you have the cash to pay for it, is one of the soundest and steadiest of investments.

希腊数学是"永恒的",它甚至比希腊文学还要长久。当埃斯库罗斯[37]被人遗忘时,阿基米德[38]仍被铭记,因为语言会消亡,而数学思想不会。"永生"也许是个愚蠢的词汇,但大概只有数学家才最有可能理解它的含义。

任何一个人也完全没有必要担心未来会对他不公平。通常,与永生相伴的是滑稽或残忍:人们不愿意成为奥格[39]、阿纳尼亚斯[40]或加利奥[41]。在数学领域,历史有时甚至也会开一些奇怪的玩笑:罗尔[42]在微积分基础课本上的形象仿佛能和牛顿比肩;法里[43]一直被提起的,是因为他没能理解一个在14年前就被哈罗斯[44]完美证明了的定理;五个值得一提的挪威人的名字被写在阿贝尔的传记里,只因他们尽心尽职的愚蠢举动伤害了自己国家最伟大的人物。但总体而言,科学史是公道的,数学史尤其如此。其他学科都没有这种明确并且能被一致认可的标准,那些被人们记住的人,几乎都是那些值得被记住的人。如果你愿意投资,数学名望一定是最合理、最稳健的选择之一。

9

All this is very comforting for dons, and especially for professors of mathematics. It is sometimes suggested, by lawyers or politicians or business men, that an academic career is one sought mainly by cautious and unambitious persons who care primarily for comfort and security. The reproach is quite misplaced. A don surrenders something, and in particular the chance of making large sums of money—it is very hard for a professor to make £2000 a year; and security of tenure is naturally one of the considerations which make this particular surrender easy. That is not why Housman would have refused to be Lord Simon or Lord Beaverbrook. He would have rejected their careers because of his ambition, because he would have scorned to be a man to be forgotten in twenty years.

Yet how painful it is to feel that, with all these advantages, one may fail. I can remember Bertrand Russell telling me of a horrible dream. He was in the top floor of the University Library, about A.D. 2100. A library assistant was going round the shelves carrying an enormous bucket, taking down book after book, glancing at them, restoring them to the shelves or dumping them into the bucket. At last he came to three large volumes which Russell could recognize as the last surviving copy of *Principia Mathematica*. He took down one of the volumes, turned over a few pages, seemed puzzled for a moment by the curious symbolism, closed the volume, balanced it in his hand and hesitated... .

9

<big>这</big>一切都让大学教师感到欣慰,尤其是数学教授。律师、政治家或商人有时会说,从事学术工作,主要是那些谨慎、没有抱负的人所追求的,他们只关心舒适和安全。这种指责是完全错误的。教师放弃了一些东西,尤其是赚大钱的机会——一位教授几乎不可能一年挣 2000 英镑。保证终身教职自然是他们考虑的因素之一,这也使他们容易做出取舍。但这并不是豪斯曼不愿成为西蒙勋爵[45]或比弗布鲁克勋爵[46]那样人的原因。他这样做,是出于他的雄心壮志,因为他不介意自己在 20 年后被人遗忘。

然而,尽管有这些好处,但考虑到有失败的可能,仍然会令人感到非常痛苦。我记得伯特兰·罗素[47]曾告诉我一个可怕的梦。他穿越到公元 2100 年左右,在大学图书馆的顶楼看到一位图书馆管理员,正提着一个大桶在书架边走来走去。管理员取下书翻阅一番,然后把它们放回书架或扔进桶里。最后,管理员走向三本厚厚的书,罗素认出它们是最后仅存的一套《数学原理》。管理员取下其中的一本,翻了几页,他似乎对这些奇特的符号系统迟疑了片刻,然后便合上书,拿在手里掂量着,犹豫着是不是……

10

A mathematician, like a painter or a poet, is a maker of patterns. If his patterns are more permanent than theirs, it is because they are made with *ideas*. A painter makes patterns with shapes and colours, a poet with words. A painting may embody an 'idea', but the idea is usually commonplace and unimportant. In poetry, ideas count for a good deal more; but, as Housman insisted, the importance of ideas in poetry is habitually exaggerated: 'I cannot satisfy myself that there are any such things as poetical ideas.... . Poetry is not the thing said but a way of saying it.'

> Not all the water in the rough rude sea Can wash the balm from an anointed King.

Could lines be better, and could ideas be at once more trite and more false? The poverty of the ideas seems hardly to affect the beauty of the verbal pattern. A mathematician, on the other hand, has no material to work with but ideas, and so his patterns are likely to last longer, since ideas wear less with time than words.

The mathematician's patterns, like the painter's or the poet's, must be *beautiful*; the ideas, like the colours or the words, must fit together in a harmonious way. Beauty is the first test: there is no permanent place in the

10

和画家或诗人一样,数学家也是模式的创造者。如果说数学家创造的模式比前者的更持久,那是因为这些模式是由思想构成的。画家用形状和颜色创造样式,诗人用文字创造格律。一幅画中可能包含"思想",但它的思想通常是老生常谈,并不怎么重要。在诗歌里,思想会更重要一些。但是,正如豪斯曼所坚持的那样,思想在诗歌中的重要性被习惯性地夸大了:"我无法说服自己,存在一种叫诗歌的思想的东西。诗歌不在于它要表达的内容,而在于它表达的方式。"

> 任怒海所有的波涛
> 　也不能涤除君王的尊荣。[48]

这诗还能更棒吗?会不会忽然觉得它的思想陈腐而不堪一击呢?思想上的匮乏似乎很难影响到语言的形式之美。然而,数学家除了思想没有别的素材,所以他们创造的模式或许更持久,因为思想比文字更经得起时间的打磨。

同画家和诗人的模式一样,数学家的模式必定是美的。与色彩和文字相同,思想也必然会以某种和谐的方式组合。美是首要的试

world for ugly mathematics. And here I must deal with a misconception which is still widespread (though probably much less so now than it was twenty years ago), what Whitehead has called the 'literary superstition' that love of and aesthetic appreciation of mathematics is 'a monomania confined to a few eccentrics in each generation'.

It would be difficult now to find an educated man quite insensitive to the aesthetic appeal of mathematics. It may be very hard to *define* mathematical beauty, but that is just as true of beauty of any kind—we may not know quite what we mean by a beautiful poem, but that does not prevent us from recognizing one when we read it. Even Professor Hogben, who is out to minimize at all costs the importance of the aesthetic element in mathematics, does not venture to deny its reality. 'There are, to be sure, individuals for whom mathematics exercises a coldly impersonal attraction... . The aesthetic appeal of mathematics may be very real for a chosen few.' But they are 'few', he suggests, and they feel 'coldly' (and are really rather ridiculous people, who live in silly little university towns sheltered from the fresh breezes of the wide open spaces). In this he is merely echoing Whitehead's 'literary superstition'.

The fact is that there are few more 'popular' subjects than mathematics. Most people have some appreciation of mathematics, just as most people can enjoy a pleasant tune; and there are probably more people really interested in mathematics than in music. Appearances may suggest the contrary, but there are easy explanations. Music can be used to stimulate mass emotion, while mathematics cannot; and musical incapacity is recognized (no doubt rightly) as mildly discreditable, whereas most people are so frightened of the name of mathematics that they are ready, quite unaffectedly, to exaggerate their own mathematical stupidity.

金石：丑陋的数学不可能永存。在这里，我必须纠正一个至今仍普遍流传的误解（尽管现在可能已经比20年前好了许多），这就是怀特海[49]所谓的"文人般的执迷"[50]，即认为对数学审美的热爱"在每一代人里都只是少数怪人的偏执"。

如今，很难找到一个受过教育的人会对数学的美学魅力无动于衷。数学之美可能很难定义，但任何一种美的定义都是如此——我们可能不太清楚一首诗的美丽是指什么，但这并不妨碍我们在阅读时感受到它。甚至连霍格本[51]教授也不敢冒风险否认这个事实，虽然他不惜一切代价地想把数学里美学元素的重要性降到最低。"可以肯定的是，数学对某些人具有一种冷峻客观的吸引力……数学的美学魅力对少数人来说可能是真实存在的。"他又提到，但这种人"很少"，并且感觉"冷冰冰的"（他们真的是相当可笑的一群人，蜗在小小的大学城里，远离外面开放世界的清新微风）。在这一点上，他只不过是在重复怀特海的"文人般的执迷"。

事实上，几乎没有比数学更"大众化"的学科。大多数人对数学有一定的鉴赏能力，就像大多数人能欣赏悦耳的曲调一样。对数学感兴趣的人可能比对音乐感兴趣的人还要多。尽管表面看起来并非如此，但这很好解释。音乐可以牵动大众的情感，但数学不能；音乐上的无感会让人觉得有些丢人（这毫无疑问是对的），然而大多数人看到数学这个词就会感到非常害怕，以至于下意识地做好了夸大自己在数学上无知的准备。

A very little reflection is enough to expose the absurdity of the 'literary superstition'. There are masses of chess-players in every civilized country—in Russia, almost the whole educated population; and every chess-player can recognize and appreciate a 'beautiful' game or problem. Yet a chess problem is *simply* an exercise in pure mathematics (a game not entirely, since psychology also plays a part), and everyone who calls a problem 'beautiful' is applauding mathematical beauty, even if it is beauty of a comparatively lowly kind. Chess problems are the hymn-tunes of mathematics.

We may learn the same lesson, at a lower level but for a wider public, from bridge, or descending further, from the puzzle columns of the popular newspapers. Nearly all their immense popularity is a tribute to the drawing power of rudimentary mathematics, and the better makers of puzzles, such as Dudeney or 'Caliban', use very little else. They know their business; what the public wants is a little intellectual 'kick', and nothing else has quite the kick of mathematics.

I might add that there is nothing in the world which pleases even famous men (and men who have used disparaging language about mathematics) quite so much as to discover, or rediscover, a genuine mathematical theorem. Herbert Spencer republished in his autobiography a theorem about circles which he proved when he was twenty (not knowing that it had been proved over two thousand years before by Plato). Professor Soddy is a more recent and a more striking example (but *his* theorem really is his own)[*].

---

[*] See his letters on the 'Hexlet' in *Nature*, vols. 137-9 (1936-7).

只要稍加反思，就足以戳穿"文人般的执迷"的荒谬。在每一个文明国家里，都有许多国际象棋棋手——比如在俄罗斯，几乎所有受过教育的人都会下棋；每位棋手都能分辨并欣赏那些"漂亮"的对弈或难题。然而，每个国际象棋难题只不过是纯数学里的一道练习题（对弈包含了心理因素，因此不完全一样），所有认为某个国际象棋难题"漂亮"的人，都是在称赞数学之美，哪怕它只是一种层次相对较低的美。国际象棋难题都是为数学写的赞美诗。

就更通俗、更广泛的程度而言，从桥牌（或者大众更加喜闻乐见的流行报纸上的字谜专栏游戏）里也能得到同样的结论。这些游戏之所以广受欢迎，几乎都仰仗基础数学的吸引力，而像杜德尼[52]或"卡利班"[53]这样的益智游戏制作高手，也很少涉及其他领域的知识。他们知道自己该做什么：公众想要的只不过是在智力上得到一点"刺激"，而数学是最合适的。

我还可以补充一点，世界上没有什么比发现或再次发现一个真正的数学定理更能让名人们（以及那些曾经大肆贬低数学的人）高兴了。赫伯特·斯宾塞[54]在他的自传中重新发表了一个关于圆的定理，那是他在20岁时证明的（他并不知道柏拉图[55]在2000多年前就已经证明过了）。而最近更著名的例子，则要算是索迪[56]教授（不过，他的定理确实是他自己发现的）。*

---

* 见他关于"六球链"的通信，《自然》杂志，137~139卷（1936~1937年）。

11

A chess problem is genuine mathematics, but it is in some way 'trivial' mathematics. However ingenious and intricate, however original and surprising the moves, there is something essential lacking. Chess problems are *unimportant*. The best mathematics is *serious* as well as beautiful—'important' if you like, but the word is very ambiguous, and 'serious' expresses what I mean much better.

I am not thinking of the 'practical' consequences of mathematics. I have to return to that point later: at present I will say only that if a chess problem is, in the crude sense, 'useless', then that is equally true of most of the best mathematics; that very little of mathematics is useful practically, and that that little is comparatively dull. The 'seriousness' of a mathematical theorem lies, not in its practical consequences, which are usually negligible, but in the *significance* of the mathematical ideas which it connects. We may say, roughly, that a mathematical idea is 'significant' if it can be connected, in a natural and illuminating way, with a large complex of other mathematical ideas. Thus a serious mathematical theorem, a theorem which connects significant ideas, is likely to lead to important advances in mathematics itself and even in other sciences. No chess problem has ever affected the general development of scientific thought; Pythagoras, Newton, Einstein have in their times changed its whole direction.

11

国际象棋难题是真正的数学，但它在某种程度上只是"平凡的"数学。无论落子多么巧妙、多么复杂、多么独特、多么奇谲，它都是先天不足的。国际象棋难题**不重要**。最好的数学不仅是美丽的，而且是**严肃的**——如果你喜欢，也可以说它是"重要的"，但这个词很含糊，而"严肃"更能表达我的意思。

我没有讨论数学的"实用"效果，这个问题我们稍后再做讨论。眼下我想说的是，粗略地说，如果一个国际象棋难题是"无用的"，那么多数最棒的数学也一样。只有微乎其微的数学知识在实践中是有用的，同时它们也比较无趣。一个数学定理的"严肃性"，是由和它相关的数学思想的**重要性**体现的，而不是它能有什么实用效果，这些效果通常可以忽略不计。大致说来，如果某个数学思想能够以一种自然而富有启发性的方式和大量其他数学思想联系起来，那么它就是"重要的"。因此，一个严肃的数学定理，即一个和重要思想有关联的定理，很可能会使数学本身，甚至其他科学取得重要的进展。没有一个国际象棋难题影响过科学的发展，但毕达哥拉斯、牛顿和爱因斯坦在他们所处的时代里，改变了整个科学发展的方向。

The seriousness of a theorem, of course, does not *lie in* its consequences, which are merely the *evidence* for its seriousness. Shakespeare had an enormous influence on the development of the English language, Otway next to none, but that is not why Shakespeare was the better poet. He was the better poet because he wrote much better poetry. The inferiority of the chess problem, like that of Otway's poetry, lies not in its consequences but in its content.

There is one more point which I shall dismiss very shortly, not because it is uninteresting but because it is difficult, and because I have no qualifications for any serious discussion in aesthetics. The beauty of a mathematical theorem *depends* a great deal on its seriousness, as even in poetry the beauty of a line may depend to some extent on the significance of the ideas which it contains. I quoted two lines of Shakespeare as an example of the sheer beauty of a verbal pattern; but

> After life's fitful fever he sleeps well

seems still more beautiful. The pattern is just as fine, and in this case the ideas have significance and the thesis is sound, so that our emotions are stirred much more deeply. The ideas do matter to the pattern, even in poetry, and much more, naturally, in mathematics; but I must not try to argue the question seriously.

定理的严肃性当然不依赖于它的结果，这些结果只不过是证明定理严肃性的证据。莎士比亚[57]对英语发展有着巨大的影响，而奥特韦[58]则几乎什么影响也没有，然而这并不是莎士比亚被认为是更优秀的诗人的原因。他之所以更优秀，是因为他写的诗更出色。就像奥特韦的诗歌一样，国际象棋难题的不足，不在于结果而在于其自身。

还有一个观点我想一笔带过，不是因为它无趣，而是因为有难度，并且我也没资格在美学方面做任何严肃的讨论。数学定理的美在很大程度上取决于它的严肃性，甚至诗歌里的诗句之美在某种程度上也是由其蕴含的思想决定的。我引用过莎士比亚的两行诗，作为语言格律的纯粹之美的例子，但是下面这行诗似乎更美：

　　　　历经癫狂人生，现睡意正酣。[59]

这行诗的格律很棒，而且思想有意义，论点也合理，正因为如此，我们的情感才会产生出更深刻的共鸣。思想对模式而言很重要（哪怕是在诗歌里），数学更是如此。但我无法严肃地讨论这个问题。

12

It will be clear by now that, if we are to have any chance of making progress, I must produce examples of 'real' mathematical theorems, theorems which every mathematician will admit to be first-rate. And here I am very heavily handicapped by the restrictions under which I am writing. On the one hand my examples must be very simple, and intelligible to a reader who has no specialized mathematical knowledge; no elaborate preliminary explanations must be needed; and a reader must be able to follow the proofs as well as the enunciations. These conditions exclude, for instance, many of the most beautiful theorems of the theory of numbers, such as Fermat's 'two square' theorem or the law of quadratic reciprocity. And on the other hand my examples should be drawn from 'pukka' mathematics, the mathematics of the working professional mathematician; and this condition excludes a good deal which it would be comparatively easy to make intelligible but which trespasses on logic and mathematical philosophy.

I can hardly do better than go back to the Greeks. I will state and prove two of the famous theorems of Greek mathematics. They are 'simple' theorems, simple both in idea and in execution, but there is no doubt at all about their being theorems of the highest class. Each is as fresh and significant as when it was discovered—two thousand years have not written a wrinkle on either of them. Finally, both the statements and the proofs can be mastered in an hour by any intelligent reader, however slender his mathematical equipment.

1. The first is Euclid's[*] proof of the existence of an infinity of prime numbers.

---

[*] *Elements* ix 20. The real origin of many theorems in the *Elements* is obscure, but there seems to be no particular reason for supposing that this one is not Euclid's own.

12

现在已经很清楚了，为了更进一步，让我来举些"真正的"数学定理的例子，并且所有数学家都认为它们是第一流的。不过，我写作的限制太多。一方面，这些例子必须非常简单，没有专门数学知识的读者也能理解，不需要额外的预备知识也能看懂解释，从而理解证明。就数论而言，这些约束排除了大多数漂亮的定理，如费马[60]"二平方"定理或二次互反律。另一方面，我举的例子应该属于"真正的"数学，它是职业数学家研究的数学。这个限制又排除了许多相对容易理解却涉及逻辑和数学哲学的东西。

于是，用古希腊的例子便是再好不过的了。下面，我会陈述并证明两个著名的古希腊数学定理。这两个定理是"简单"的，无论其思想还是演算都是如此，但它们毫无疑问也都是最高层次的定理。如今，它们的新鲜度和重要性和它们刚被发现时相比丝毫不减——两千年的时光没有对它们留下任何痕迹。最后提一句，就算没有数学基础，聪明的读者也都能在一小时内掌握这些定理和它们的证明。

1. 欧几里得[61]关于存在无穷多个素数的证明*。

---

* 《几何原本》第9章第20节。《几何原本》里的许多定理的真正源头都很含糊，但似乎没什么明确的证据支持这个定理并非出自欧几里得本人。

The *prime numbers* or *primes* are the numbers

(A)　　　2, 3, 5, 7, 11, 13, 17, 19, 23, 29, ⋯

which cannot be resolved into smaller factors*. Thus 37 and 317 are prime. The primes are the material out of which all numbers are built up by multiplication: thus $666 = 2 \cdot 3 \cdot 3 \cdot 37$. Every number which is not prime itself is divisible by at least one prime (usually, of course, by several). We have to prove that there are infinitely many primes, i.e. that the series (A) never comes to an end.

Let us suppose that it does, and that

$$2, 3, 5, \cdots, P$$

is the complete series (so that P is the largest prime); and let us, on this hypothesis, consider the number Q defined by the formula

$$Q = (2 \cdot 3 \cdot 5 \cdot \cdots \cdot P) + 1.$$

It is plain that Q is not divisible by any of 2, 3, 5,⋯, P; for it leaves the remainder 1 when divided by any one of these numbers. But, if not itself prime, it is divisible by *some* prime, and therefore there is a prime (which may be Q itself) greater than any of them. This contradicts our hypothesis, that there is no prime greater than P; and therefore this hypothesis is false.

The proof is by *reductio ad absurdum*, and *reductio ad absurdum*, which Euclid loved so much, is one of a mathematician's finest weapons**. It is a far finer gambit than any chess gambit: a chess player may offer the sacrifice of a pawn or even a piece, but a mathematician offers *the game*.

---

* There are technical reasons for not counting 1 as a prime.
** The proof can be arranged so as to avoid a *reductio*, and logicians of some schools would prefer that it should be.

素数就是下面这样的数

$$2, 3, 5, 7, 11, 13, 17, 19, 23, 29, \cdots \quad (A)$$

它们不能被分解成更小的因数[*]。因此，37 和 317 是素数。所有整数都是由素数相乘而得到的，如 $666=2\times 3\times 3\times 37$。每个不是素数的整数都能被至少一个（当然，通常会是若干个）素数整除。我们必须证明有无穷多个素数，即数列 (A) 永远不会终止。

让我们先假设数列 (A) 会终止，也就是说

$$2, 3, 5, \cdots, P$$

是完整的素数数列（因此 P 是最大的素数）。在这个假设下，我们考虑由以下公式定义的整数 Q:

$$Q = (2\times 3\times 5\times \cdots \times P)+1$$

显然，Q 不能被 $2, 3, 5, \cdots, P$ 中的任何一个整除，因为 Q 除以这些数中的任何一个都会余 1。但是，如果它本身不是素数，那么它就可以被某个素数整除，因此存在一个素数（可能是 Q 本身），它比数列 (A) 里的任何一个素数都大。这和我们"不存在大于 P 的素数"的假设矛盾，因此原假设错误。

证明用的是**反证法**，欧几里得非常喜爱这种方法，它是数学家最擅长的武器之一[**]。这一招比象棋中任何的弃子开局法都要高明：棋手可能会舍弃一个兵或其他棋子，而数学家舍掉的是**整个棋局**。

---

[*] 出于某些技术原因，我们不把 1 当作素数。
[**] 不用反证法也可以得到证明，某些逻辑学派更愿意这样做（不用反证法）。

13

2. My second example is Pythagoras's[*] proof of the 'irrationality' of $\sqrt{2}$.

A 'rational number' is a fraction a/b, where *a* and *b* are integers; we may suppose that *a* and *b* have no common factor, since if they had we could remove it. To say that '$\sqrt{2}$ is irrational' is merely another way of saying that 2 cannot be expressed in the form $(a/b)^2$; and this is the same thing as saying that the equation

(B) $$a^2 = 2b^2$$

cannot be satisfied by integral values of *a* and *b* which have no common factor. This is a theorem of pure arithmetic, which does not demand any knowledge of 'irrational numbers' or depend on any theory about their nature.

We argue again by *reductio ad absurdum*; we suppose that (B) is true, *a* and *b* being integers without any common factor. It follows from (B) that a^2 is even (since $2b^2$ is divisible by 2), and therefore that *a* is even (since the square of an odd number is odd). If *a* is even then

(C) $$a = 2c$$

for some integral value of *c*; and therefore

---

[*] The proof traditionally ascribed to Pythagoras, and certainly a product of his school. The theorem occurs, in a much more general form, in Euclid (*Elements* x 9).

13

2\. 我的第二个例子,是毕达哥拉斯关于 $\sqrt{2}$ 是无理数的证明[*]。

所谓"有理数"是指分数 a/b,其中 a 和 b 都是整数。我们假设 a 和 b 没有公因数,倘若它们有公因数,那么可以把它约掉。"$\sqrt{2}$ 是无理数"只不过是"2 不能表示为 $(a/b)^2$ 的形式"的另一种说法。也就是说,不存在没有公因数的整数 a 和 b 可以满足方程

$$a^2 = 2b^2 \qquad (B)$$

这是一个纯算术定理,它不需要任何关于"无理数"的知识,也不依赖于和无理数性质有关的理论。

我们还是用**反证法**来证明:假设 (B) 为真,a 和 b 是没有公因数的整数。由 (B) 可以推出,a^2 是偶数(因为 $2b^2$ 可以被 2 整除),因此 a 也是偶数(因为奇数的平方是奇数)。如果 a 是偶数,那么对某个整数 c 有

$$a = 2c \qquad (C)$$

因此,

---

[*] 这个证明传统上归功于毕达哥拉斯,可以肯定这是毕达哥拉斯学派的成果。这个定理更一般的形式见《几何原本》第 10 章第 9 节。

$$2b^2 = a^2 = (2c)^2 = 4c^2$$

or

(D) $$b^2 = 2c^2$$

Hence b^2 is even, and therefore (for the same reason as before) b is even. That is to say, a and b are both even, and so have the common factor 2. This contradicts our hypothesis, and therefore the hypothesis is false.

It follows from Pythagoras's theorem that the diagonal of a square is incommensurable with the side (that their ratio is not a rational number, that there is no unit of which both are integral multiples). For if we take the side as our unit of length, and the length of the diagonal is d, then, by a very familiar theorem also ascribed to Pythagoras[*],

$$d^2 = 1^2 + 1^2 = 2$$

so that d cannot be a rational number.

I could quote any number of fine theorems from the theory of numbers whose *meaning* anyone can understand. For example, there is what is called 'the fundamental theorem of arithmetic', that any integer can be resolved, *in one way only*, into a product of primes. Thus $666 = 2 \cdot 3 \cdot 3 \cdot 37$, and there is no other decomposition; it is impossible that $666 = 2 \cdot 11 \cdot 29$ or that $13 \cdot 89 = 17 \cdot 73$ (and we can see so without working out the products). This theorem is, as its name implies, the foundation of higher arithmetic; but the proof, although not 'difficult', requires a certain amount of preface and might be found tedious by an unmathematical reader.

---

[*] Euclid, *Elements* 1 47.

$$2b^2 = a^2 = (2c)^2 = 4c^2$$

于是得到

$$b^2 = 2c^2 \qquad (D)$$

因此 b^2 是偶数，b 也是偶数（理由如前）。也就是说，a 和 b 都是偶数，所以它们有公因数 2。这与我们的假设矛盾，因此原假设错误。

由毕达哥拉斯定理可知，正方形的对角线与边是不可公度的（它们的比值不是有理数，即不存在一个公共的单位，使它们同为其整数倍）。如果我们把正方形的边取为一个长度单位，令对角线的长度为 d，那么，根据一个我们非常熟悉的定理[62]，它也归功于毕达哥拉斯*，可以得到

$$d^2 = 1^2 + 1^2 = 2$$

所以 d 不可能是有理数。

我可以从数论里找到许多优秀的定理，任何人都能理解它们的含义。例如，有一个所谓的"算术基本定理"，即任何整数都只能以一种方式分解为素数的乘积。因此 $666 = 2 \times 3 \times 3 \times 37$，它没有其他的分解方式。不可能有 $666 = 2 \times 11 \times 29$，也不可能有 $13 \times 89 = 17 \times 73$（不用算出乘积，我们就能知道）。正如这个定理的名字所暗示的，它是高等算术的基础。然而，证明它虽然并不"困难"，但需要一些铺垫，对非数学专业的读者而言，可能会感觉很乏味。

---

* 见欧几里得的《几何原本》第 1 章第 47 节。

Another famous and beautiful theorem is Fermat's 'two square' theorem. The primes may (if we ignore the special prime 2) be arranged in two classes; the primes

$$5, 13, 17, 29, 37, 41, \cdots$$

which leave remainder 1 when divided by 4, and the primes

$$3, 7, 11, 19, 23, 31, \cdots$$

which leave remainder 3. All the primes of the first class, and none of the second, can be expressed as the sum of two integral squares: thus

$$5 = 1^2 + 2^2, \ 13 = 2^2 + 3^2$$
$$17 = 1^2 + 4^2, \ 29 = 2^2 + 5^2$$

but 3, 7, 11, and 19 are not expressible in this way (as the reader may check by trial). This is Fermat's theorem, which is ranked, very justly, as one of the finest of arithmetic. Unfortunately there is no proof within the comprehension of anybody but a fairly expert mathematician.

There are also beautiful theorems in the 'theory of aggregates' (*Mengenlehre*), such as Cantor's theorem of the 'non-enumerability' of the continuum. Here there is just the opposite difficulty. The proof is easy enough, when once the language has been mastered, but considerable explanation is necessary before the *meaning* of the theorem becomes clear. So I will not try to give more examples. Those which I have given are test cases, and a reader who cannot appreciate them is unlikely to appreciate anything in mathematics.

I said that a mathematician was a maker of patterns of ideas, and that beauty and seriousness were the criteria by which his patterns should be judged. I can hardly believe that anyone who has understood the two theorems will dispute that they pass these tests. If we compare them

另一个著名的优美定理是费马"二平方"定理。如果不考虑特殊的素数 2，素数可以分成两类。第一类素数

$$5, 13, 17, 29, 37, 41, \cdots$$

除以 4 的余数是 1，第二类素数

$$3, 7, 11, 19, 23, 31, \cdots$$

除以 4 的余数是 3。所有第一类素数都能表示成两个整数的平方和，而所有第二类素数都无法如此表示。例如，

$$5 = 1^2 + 2^2, \quad 13 = 2^2 + 3^2$$
$$17 = 1^2 + 4^2, \quad 29 = 2^2 + 5^2$$

但是 3, 7, 11, 19 则没有这种表示形式（读者可以试着检验一下）。这就是费马"二平方"定理，非常公正地说，这可列为算术中最好的结果之一。遗憾的是，还没有人人都能看得懂的证明方法，只有具有相当数学知识的人，才能看明白那些证明。[63]

在"集合论"里也有一些优美的定理，比如康托尔[64]关于连续统"不可数"的定理。这个定理遇到的麻烦正好相反。一旦掌握了它的语言，证明起来会很容易，但在搞清楚定理的含义之前，还需要大量解释。因此，我就不再多举例子了。我所举的例子都可以作为测试，如果读者无法欣赏它们，那么可能也欣赏不了数学。

我说过，数学家是思想模式的创造者，美丽和严肃性是判断其模式的标准。我无法相信，任何一个理解这两个定理的人会怀疑它们不符合这些标准。如果我们将这两个定理与杜德尼构思的最巧的

with Dudeney's most ingenious puzzles, or the finest chess problems that masters of that art have composed, their superiority in both respects stands out: there is an unmistakable difference of class. They are much more serious, and also much more beautiful; can we define, a little more closely, where their superiority lies?

谜题（或者国际象棋大师编制的最妙的难题）相比，那么它们在这两个方面的优势都很明显：毫无疑问，它们完全不在一个层次。定理严肃得多，也漂亮得多。我们还能更进一步说出它们到底有什么优势吗？

14

In the first place, the superiority of the mathematical theorems in *seriousness* is obvious and overwhelming. The chess problem is the product of an ingenious but very limited complex of ideas, which do not differ from one another very fundamentally and have no external repercussions. We should think in the same way if chess had never been invented, whereas the theorems of Euclid and Pythagoras have influenced thought profoundly, even outside mathematics.

Thus Euclid's theorem is vital for the whole structure of arithmetic. The primes are the raw material out of which we have to build arithmetic, and Euclid's theorem assures us that we have plenty of material for the task. But the theorem of Pythagoras has wider applications and provides a better text.

We should observe first that Pythagoras's argument is capable of far-reaching extension, and can be applied, with little change of principle, to very wide classes of 'irrationals'. We can prove very similarly (as Theodorus seems to have done) that

$$\sqrt{3},\ \sqrt{5},\ \sqrt{7},\ \sqrt{11},\ \sqrt{13},\ \sqrt{17}$$

are irrational, or (going beyond Theodorus) that $\sqrt[3]{2}$ and $\sqrt[3]{7}$ are irrational[*].

---

[*] See Ch. IV of Hardy and Wright's *Introduction to the Theory of Numbers*, where there are discussions of different generalizations of Pythagoras's argument, and of a historical puzzle about Theodorus.

14

首先,数学定理在严肃性上的优越性是显而易见的,也是毫无争议的。国际象棋难题虽然巧妙,但其思想的复杂性非常有限,它们的思路没有本质区别,也不会对外界产生什么影响。即便国际象棋从未被发明过,我们应该也会以同样的方式思考,而欧几里得定理和毕达哥拉斯定理即使在数学领域之外,都深刻地影响着我们的思想。

因此,欧几里得定理[65]对于整个算术结构是至关重要的。素数是算术的基础,欧几里得定理保证我们有足够多的素数可以用于构造算术。不过,毕达哥拉斯定理有着更广泛的应用,示范性也更强。

我们首先应该注意到,毕达哥拉斯的论证具有深远的推广意义,只要稍作改变,它就可以被应用到广阔的"无理数"领域。我们可以用非常类似的方法(就像西奥多罗斯[66]那样)证明

$$\sqrt{3}, \sqrt{5}, \sqrt{7}, \sqrt{11}, \sqrt{13}, \sqrt{17}$$

是无理数,或者(超越西奥多罗斯)证明 $\sqrt[3]{2}$ 和 $\sqrt[3]{7}$ 也是无理数[*]。

---

[*] 参见哈代和赖特合著的《哈代数论(第6版)》第4章,其中讨论了毕达哥拉斯论证法的各种泛化,以及西奥多罗斯的历史之谜。

Euclid's theorem tells us that we have a good supply of material for the construction of a coherent arithmetic of the integers. Pythagoras's theorem and its extensions tell us that, when we have constructed this arithmetic, it will not prove sufficient for our needs, since there will be many magnitudes which obtrude themselves upon our attention and which it will be unable to measure; the diagonal of the square is merely the most obvious example. The profound importance of this discovery was recognized at once by the Greek mathematicians. They had begun by assuming (in accordance, I suppose, with the 'natural' dictates of 'common sense') that all magnitudes of the same kind are commensurable, that any two lengths, for example, are multiples of some common unit, and they had constructed a theory of proportion based on this assumption. Pythagoras's discovery exposed the unsoundness of this foundation, and led to the construction of the much more profound theory of Eudoxus which is set out in the fifth book of the *Elements*, and which is regarded by many modern mathematicians as the finest achievement of Greek mathematics. This theory is astonishingly modern in spirit, and may be regarded as the beginning of the modern theory of irrational number, which has revolutionized mathematical analysis and had much influence on recent philosophy.

There is no doubt at all, then, of the 'seriousness' of either theorem. It is therefore the better worth remarking that neither theorem has the slightest 'practical' importance. In practical applications we are concerned only with comparatively small numbers; only stellar astronomy and atomic physics deal with 'large' numbers, and they have very little more practical importance, as yet, than the most abstract pure mathematics. I do not know what is the highest degree of accuracy which is ever useful to an engineer—we shall be very generous if we say ten significant figures. Then

$$3.141, 592, 654$$

根据欧几里得定理可知，我们有足够多的素数可以用于构造清晰的整数算术。毕达哥拉斯定理及其推广则告诉我们，当完成这种算术的构造后，并不能证明它已经满足了我们的需求，因为有许多"量"迫使我们去关注，它们是无法度量的：正方形的对角线只不过是一个最明显的例子。古希腊数学家很快就意识到了这个发现的深远意义。一开始，他们假设（我认为是根据"常识"的"自然"要求）所有同类的量都是可公度的，例如任何两个长度都是某个公共单位的倍数，并且他们基于这个假设构建了一套比例理论。毕达哥拉斯的发现揭示了这个基础并不牢靠，它导致了更为深奥的欧多克索斯[67]理论的建立，这一理论出现在《几何原本》第5章，许多现代数学家认为它是古希腊数学最杰出的成就。这个理论具有惊人的现代性，可以说是现代无理数理论的开端。无理数理论彻底改变了数学分析，也深刻影响了近代哲学。

因此，这两个定理中的任何一个的"严肃性"都是无可争辩的。更值得注意的是，这两个定理都没有丝毫的"实际"意义。在实际应用中我们只关心相对较小的数，只有恒星天文学和原子物理学会处理"大"数。迄今为止，它们在实用性方面几乎也不比最抽象的纯数学更重要。我不知道对工程师而言，有用的最高精度是多少——如果我们用10位有效数字，那应该是绰绰有余了。比如

$$3.141\,592\,654$$

(the value of π to nine places of decimals) is the ratio

$$\frac{3{,}141{,}592{,}654}{1{,}000{,}000{,}000}$$

of two numbers of ten digits. The number of primes less than 1,000,000,000 is 50,847,478: that is enough for an engineer, and he can be perfectly happy without the rest. So much for Euclid's theorem; and, asregards Pythagoras's, it is obvious that irrationals are uninteresting to an engineer, since he is concerned only with approximations, and all approximations are rational.

（π 精确到小数点后 9 位的值）就是两个 10 位数的比值：

$$\frac{3\ 141\ 592\ 654}{1\ 000\ 000\ 000}$$

小于 1 000 000 000 的素数有 50 847 478 个，这对一个工程师来说，已经足够了，倘若没有其他素数，他也可以过得非常快乐。欧几里得定理就此告一段落。至于毕达哥拉斯定理，工程师显然对无理数并不感兴趣，因为他只关心近似值，而所有近似值都是有理数。

15

A 'serious' theorem is a theorem which contains 'significant' ideas, and I suppose that I ought to try to analyse a little more closely the qualities which make a mathematical idea significant. This is very difficult, and it is unlikely that any analysis which I can give will be very valuable. We can recognize a 'significant' idea when we see it, as we can those which occur in my two standard theorems; but this power of recognition requires a rather high degree of mathematical sophistication, and of that familiarity with mathematical ideas which comes only from many years spent in their company. So I must attempt some sort of analysis; and it should be possible to make one which, however inadequate, is sound and intelligible so far as it goes. There are two things at any rate which seem essential, a certain *generality* and a certain *depth*; but neither quality is easy to define at all precisely.

A significant mathematical idea, a serious mathematical theorem, should be 'general' in some such sense as this. The idea should be one which is a constituent in many mathematical constructs, which is used in the proof of theorems of many different kinds. The theorem should be one which, even if stated originally (like Pythagoras's theorem) in a quite special form, is capable of considerable extension and is typical of a whole class of theorems of its kind. The relations revealed by the proof should be such as connect many different mathematical ideas.

15

所谓"严肃"的定理是那种包含了"重要"思想的定理，我想我应该更细致地分析一下，是什么使一个数学思想意义变得重大。这是非常困难的，我所能给出的任何分析，其实都未必非常有价值。当看到一个"重要"的概念时，我们可以辨识出它，就像在前面的两个基本定理中提到的那些概念一样。但这种辨别能力不但需要很高的数学修养，而且要很熟悉数学思想，而这种熟悉程度只有在从事多年相关工作后才能获得。所以我必须尝试做出一些分析，尽管会不完美，但仍可以做到尽量合理且易于理解。无论从哪个角度衡量，有两件事似乎是不可或缺的：它们有一定程度的**普遍性**和**深度**，但都不太容易精确定义。

一个重要的数学思想、一个严肃的数学定理，在某种意义上应该是"普遍的"。这个数学思想应该是许多数学结构的组成部分，它们被用于证明各种不同类型的定理。这种定理应该是这样的：即便它最初被表述成一种非常特殊的形式（如毕达哥拉斯定理），但其具有相当大的可拓展性，并且是所有同类型定理的典型代表。证明所揭示的，是那种把许多不同的数学思想联系起来的关系。所有这些

All this is very vague, and subject to many reservations. But it is easy enough to see that a theorem is unlikely to be serious when it lacks these qualities conspicuously; we have only to take examples from the isolated curiosities in which arithmetic abounds. I take two, almost at random, from Rouse Ball's *Mathematical Recreations*[*].

(a) 8712 and 9801 are the only four-figure numbers which are integral multiples of their 'reversals':

$$8712 = 4 \cdot 2178, 9801 = 9 \cdot 1089$$

and there are no other numbers below 10,000 which have this property.

(b) There are just four numbers (after 1) which are the sums of the cubes of their digits, viz.

$$153 = 1^3 + 5^3 + 3^3, 370 = 3^3 + 7^3 + 0^3,$$
$$371 = 3^3 + 7^3 + 1^3, 407 = 4^3 + 0^3 + 7^3.$$

These are odd facts, very suitable for puzzle columns and likely to amuse amateurs, but there is nothing in them which appeals much to a mathematician. The proofs are neither difficult nor interesting—merely a little tiresome. The theorems are not serious; and it is plain that one reason (though perhaps not the most important) is the extreme speciality of both the enunciations and the proofs, which are not capable of any significant generalization.

---

[*] 11th edition, 1939 (revised by H. S. M. Coxeter).

都非常含糊,并且有所保留。但很容易看出,如果一个定理一点儿这些特性都没有,那么它就不太可能是严肃的;这类例证可以从众多孤立的奇特算术中找到。在劳斯·鲍尔[68]的《数学娱乐》*[69]里,我随手就能找到两个这样的例子。

(a) 在 4 位数里,只有 8712 和 9801 是其"反转数"的整数[70]倍:
$$8712 = 4 \times 2178, 9801 = 9 \times 1089$$
在小于 10 000 的数中,没有其他数具有这个性质。

(b) 在 1 之后,只有 4 个数等于它各位数字的立方和。
$$153 = 1^3 + 5^3 + 3^3, 370 = 3^3 + 7^3 + 0^3$$
$$371 = 3^3 + 7^3 + 1^3, 407 = 4^3 + 0^3 + 7^3$$

这些奇特的事实非常适合谜题专栏,业余爱好者可能会感兴趣,但它们没有吸引数学家的地方。相关的证明既不困难也不有趣——只有些许无聊。这些定理并不严肃,很明显,其中一个原因(尽管它可能不是最重要的)是这些定理和证明都是极端特殊的,人们无法对其进行任何有意义的普遍化。

---

* 见 1939 年的第 11 版(由 H. S. M. 考克斯特修订)。

16

'GENERALITY' is an ambiguous and rather dangerous word, and we must be careful not to allow it to dominate our discussion too much. It is used in various senses both in mathematics and in writings about mathematics, and there is one of these in particular, on which logicians have very properly laid great stress, which is entirely irrelevant here. In this sense, which is quite easy to define, *all* mathematical theorems are equally and completely 'general'.

'The certainty of mathematics', says Whitehead*, 'depends on its complete abstract generality.' When we assert that 2 + 3 = 5, we are asserting a relation between three groups of 'things'; and these 'things' are not apples or pennies, or things of any one particular sort or another, but *just* things, 'any old things'. The meaning of the statement is entirely independent of the individualities of the members of the groups. All mathematical 'objects' or 'entities' or 'relations', such as '2', '3', '5', ' + ', or ' = ', and all mathematical propositions in which they occur, are completely general in the sense of being completely abstract. Indeed one of Whitehead's words is superfluous, since generality, in this sense, *is* abstractness.

This sense of the word is important, and the logicians are quite right to stress it, since it embodies a truism which a good many people

---

* *Science and the Modern World*, p.33.

16

"**普**遍性"是一个模棱两可而且相当危险的词,我们必须小心不要让它过多地主导我们的讨论。它在数学以及和数学有关的著作中有许多不同的含义,其中有一种是逻辑学家特别强调的,我们在这里不做讨论。除去那种,它是很容易被定义的,所有的数学定理都具有同等且彻底的普遍性。

怀特海说:"数学的确定性是由其完全抽象的普遍性决定的。"*[71] 当我们断言 2 + 3 = 5 时,我们指的是 3 组"东西"之间的关系。这些"东西"既不是苹果也不是便士,也不是任何一种特殊东西,它只是"东西",随便是什么。这个断言的含义和群体成员的个体特征毫无关系。所有的数学"对象""实体"或"关系",如"2""3""5""+""=",以及所有出现了它们的数学命题,在完全抽象的意义上都是绝对普遍的。事实上,怀特海的用词稍显多余,因为从这个意义上讲,普遍性就是抽象性。

这个词的意义很重要,逻辑学家强调它是非常正确的,因为它蕴含了一个大多数人应该知道但又容易忘记的老生常谈。例如,某

---

* 《科学与近代世界》,第 33 页。

who ought to know better are apt to forget. It is quite common, for example, for an astronomer or a physicist to claim that he has found a 'mathematical proof' that the physical universe must behave in a particular way. All such claims, if interpreted literally, are strictly nonsense. It *cannot* be possible to prove mathematically that there will be an eclipse to-morrow, because eclipses, and other physical phenomena, do not form part of the abstract world of mathematics; and this, I suppose, all astronomers would admit when pressed, however many eclipses they may have predicted correctly.

It is obvious that we are not concerned with this sort of 'generality' now. We are looking for *differences* of generality between one mathematical theorem and another, and in Whitehead's sense all are equally general. Thus the 'trivial' theorems (*a*) and (*b*) of §15 are just as 'abstract' or 'general' as those of Euclid and Pythagoras, and so is a chess problem. It makes no difference to a chess problem whether the pieces are white and black, or red and green, or whether there are physical 'pieces' at all; it is the *same* problem which an expert carries easily in his head and which we have to reconstruct laboriously with the aid of the board. The board and the pieces are mere devices to stimulate our sluggish imaginations, and are no more essential to the problem than the blackboard and the chalk are to the theorems in a mathematical lecture.

It is not this kind of generality, common to all mathematical theorems, which we are looking for now, but the more subtle and elusive kind of generality which I tried to describe in rough terms in §15. And we must be careful not to lay *too* much stress even on generality of this kind (as I think logicians like Whitehead tend to do). It is not mere 'piling of subtlety of generalization upon subtlety of generalization'[*] which is the outstanding achievement of modern mathematics. Some measure

---

[*] *Science and the Modern World*, p.44.

位天文学家或物理学家声称他发现了一个"数学证明",证明物质的宇宙按特定规律运行,这种情况很常见。如果从字面上解释,所有这种声明都是一派胡言。从数学上证明明天会发生日食是不可能的,因为日食和其他物理现象都不是抽象数学世界的一部分。我想,所有的天文学家都不得不承认这点,无论他们成功预测过多少次日食。

很明显,在这里我们并不关心这种"普遍性"。我们讨论的是不同数学定理之间关于普遍性的差异,对怀特海而言,它们的普遍性是一样的。因此,第15节里提到的"平凡的"定理(a)和(b),在"抽象性"及"普遍性"方面,与欧几里得定理或毕达哥拉斯定理甚至国际象棋难题都是一样的。棋子是白是黑,是红是绿,或者是否存在物理上的"棋子",对国际象棋难题来说没什么区别。这些轻松记在象棋大师脑子里的难题,和我们费劲地在棋盘上摆出来的棋局是一样的。棋盘和棋子仅仅是刺激我们迟钝的想象力的道具,它们对国际象棋难题的重要性不亚于黑板和粉笔在数学讲座上对数学定理的重要性。

这种普遍性和我们如今探索的数学定理里的普遍性不同,它更像是我在第15节里大致讨论的那种微妙而又难以捉摸的普遍性。我们必须小心,不要过分强调这种普遍性(我认为怀特海这样的逻辑学家更倾向如此)。它不仅仅是"把普遍性的微妙之处相互叠加"[*],后者是现代数学的杰出成果。任何一个高级定理都必须具备某种衡

---

[*] 《科学与近代世界》,第44页。

of generality must be present in any high-class theorem, but *too much* tends inevitably to insipidity. 'Everything is what it is, and not another thing', and the differences between things are quite as interesting as their resemblances. We do not choose our friends because they embody all the pleasant qualities of humanity, but because they are the people that they are. And so in mathematics; a property common to too many objects can hardly be very exciting, and mathematical ideas also become dim unless they have plenty of individuality. Here at any rate I can quote Whitehead on my side: 'it is the large generalization, limited by a happy particularity, which is the fruitful conception[*].'

---

[*] *Science and the Modern World*, p.46.

量普遍性的标准，但普遍性过甚又必然流于乏味。"每件事都有它原本的模样，而不是另一件事"，事物之间的差异和相似之处同样有趣。我们交朋友，不是因为他们表现出人类的全部优秀品质，而是因为他们就是他们自己。数学也是如此，有共同属性的对象太多会很难让人兴奋起来，数学思想也会变得暗淡无光，除非它们有足够的个体特征。至少，我可以引用怀特海的话："只有被合适的条件限制的广泛普遍性，才能成为卓有成效的概念。"*

---

* 《科学与近代世界》，第46页。

17

The second quality which I demanded in a significant idea was *depth*, and this is still more difficult to define. It has *something* to do with *difficulty*; the 'deeper' ideas are usually the harder to grasp: but it is not at all the same. The ideas underlying Pythagoras's theorem and its generalizations are quite deep, but no mathematician now would find them difficult. On the other hand a theorem may be essentially superficial and yet quite difficult to prove (as are many 'Diophantine' theorems, i.e. theorems about the solution of equations in integers).

It seems that mathematical ideas are arranged somehow in strata, the ideas in each stratum being linked by a complex of relations both among themselves and with those above and below. The lower the stratum, the deeper (and in general the more difficult) the idea. Thus the idea of an 'irrational' is deeper than that of an integer; and Pythagoras's theorem is, for that reason, deeper than Euclid's.

Let us concentrate our attention on the relations between the integers, or some other group of objects lying in some particular stratum. Then it may happen that one of these relations can be comprehended completely, that we can recognize and prove, for example, some property of the integers, without any knowledge of the contents of lower strata. Thus we proved Euclid's theorem by consideration of properties of integers only. But there are also many theorems about integers which we cannot appreciate properly, and still less prove, without digging deeper and considering what happens below.

17

我对"重要思想"的第二个要求是**深度**,这一点更难定义。它和**难度**有关,"更深刻"的思想通常更难理解,但它们并不完全相同。毕达哥拉斯定理的基本思想及其推广是相当深刻的,但现在的数学家都不会认为它们很难。另外,某个定理可能本质上很浅显,却很难证明(就像许多"丢番图[72]的"定理一样,所谓丢番图定理是一些关于方程的整数解的定理)。

数学思想似乎是分层排列的,每一层的思想之间由一种复杂的关系相连,上下层之间也互有联系。层次越往下,思想越深刻(通常也会更难)。因此,"无理数"的概念比整数的概念更深奥,而毕达哥拉斯定理也比欧几里得定理更深刻。

让我们把注意力集中到整数,或者某一特定层次里的某组对象之间的关系上。那么,我们也许可以完全理解其中的某种关系,例如,我们可以发现和证明整数的某些性质,而无须掌握下一层的知识。因此,我们只用整数的性质就证明了欧几里得定理。但还有许多关于整数的定理,如果我们不深入研究和思考下一层的情况,就无法正确理解它们,更谈不上证明。

It is easy to find examples in the theory of prime numbers. Euclid's theorem is very important, but not very deep: we can prove that there are infinitely many primes without using any notion deeper than that of 'divisibility'. But new questions suggest themselves as soon as we know the answer to this one. There is an infinity of primes, but how is this infinity distributed? Given a large number N, say 10^{80} or $10^{10^{10}}$,* about how many primes are there less than N?** When we ask *these* questions, we find ourselves in a quite different position. We can answer them, with rather surprising accuracy, but only by boring much deeper, leaving the integers above us for a while, and using the most powerful weapons of the modern theory of functions. Thus the theorem which answers our questions (the so-called 'Prime Number Theorem') is a much deeper theorem than Euclid's or even Pythagoras's.

I could multiply examples, but this notion of 'depth' is an elusive one even for a mathematician who can recognize it, and I can hardly suppose that I could say anything more about it here which would be of much help to other readers.

---

* It is supposed that the number of protons in the universe is about 10^{80}. The number $10^{10^{10}}$, if written at length, would occupy about 50,000 volumes of average size.

** As I mentioned in §14, there are 50,847,478 primes less than 1,000,000,000; but that is as far as our *exact* knowledge extends.

在素数理论里很容易找到这样的例子。欧几里得定理很重要，但不是很深刻：我们可以不用比"可除性"更深奥的概念，就证明素数有无穷多个。然而，当我们知道了这个问题的答案后，新的问题又出现了。素数有无穷多个，但这无穷多个是如何分布的呢？给定一个很大的数 N，比方说 10^{80*} 或 $10^{10^{10}**}$，小于 N 的素数又有多少个呢？当提出这些问题的时候，我们发现自己处于完全不同的境况。我们可以用令人吃惊的准确度来回答它们，但只能暂时把整数放一边，用现代函数论中最强大的武器进行更深入的研究。因此，解答这个问题的定理（即所谓的"素数定理"）比欧几里得定理甚至毕达哥拉斯定理都要深刻得多。

我可以举出很多例子，然而对于数学家而言，即使他们能辨识出"深度"，这个概念仍旧是难以捉摸的。我也很难想象自己还能再说些什么才能对读者有所帮助。

---

* 据推测，整个宇宙的质子数量大约为 10^{80}。大约要用 50 000 本普通厚度的书，才能把 $10^{10^{10}}$ 写下来。

** 正如第 14 节所说的，小于 1 000 000 000 的素数有 50 847 478 个，但这是我们所知道的确切信息的极限了。

18

There is still one point remaining over from §11, where I started the comparison between 'real mathematics' and chess. We may take it for granted now that in substance, seriousness, significance, the advantage of the real mathematical theorem is overwhelming. It is almost equally obvious, to a trained intelligence, that it has a great advantage in beauty also; but this advantage is much harder to define or locate, since the *main* defect of the chess problem is plainly its 'triviality', and the contrast in this respect mingles with and disturbs any more purely aesthetic judgement. What 'purely aesthetic' qualities can we distinguish in such theorems as Euclid's and Pythagoras's? I will not risk more than a few disjointed remarks.

In both theorems (and in the theorems, of course, I include the proofs) there is a very high degree of *unexpectedness*, combined with *inevitability* and *economy*. The arguments take so odd and surprising a form; the weapons used seem so childishly simple when compared with the far-reaching results; but there is no escape from the conclusions. There are no complications of detail—one line of attack is enough in each case; and this is true too of the proofs of many much more difficult theorems, the full appreciation of which demands quite a high degree of technical proficiency. We do not want many 'variations' in the proof of a mathematical theorem: 'enumeration of cases', indeed, is one of the duller forms of mathematical

18

我在第 11 节开始对"真正的数学"和国际象棋做比较的时候，还留了一个问题。如今，我们可以理所当然地认为，在实质性、严肃性和重要性方面，真正的数学定理具有压倒性的优势。对受过良好教育的人而言，几乎同样显而易见的是，数学在美的方面也有很大优势，但这种优势更难定义和定位。国际象棋难题的**主要**不足显然是它太过"平凡"了，因此这方面的对比混淆和干扰了更纯粹的审美判断。在欧几里得定理和毕达哥拉斯定理中，我们能区分出哪些是"纯粹的审美"吗？我不会再冒险表达一些毫无章法的评论。

这两个定理（当然，这里的定理也包括证明）除了**必然和简练**之外，都很出人意料。它们的论证方法令人感到非常古怪和惊讶：相较于它们影响深远的结果，其所使用的论证方法是那么幼稚简单，然而得到的结论却又毫无争议。证明的细节并不复杂——每个证明都只需要从一个点切入就足够了。很多更难的定理的证明也是如此，充分理解这些定理需要相当高的技术熟练度。我们不希望在数学定理的证明里有太多"变化"——"枚举法"着实是数学论证中比较

argument. A mathematical proof should resemble a simple and clear-cut constellation, not a scattered cluster in the Milky Way.

A chess problem also has unexpectedness, and a certain economy; it is essential that the moves should be surprising, and that every piece on the board should play its part. But the aesthetic effect is cumulative. It is essential also (unless the problem is too simple to be really amusing) that the key-move should be followed by a good many variations, each requiring its own individual answer. 'If P-B5 then Kt-R6; if ... then ...; if ... then ...'— the effect would be spoilt if there were not a good many different replies. All this is quite genuine mathematics, and has its merits; but it is just that 'proof by enumeration of cases' (and of cases which do not, at bottom, differ at all profoundly[*]) which a real mathematician tends to despise.

I am inclined to think that I could reinforce my argument by appealing to the feelings of chess-players themselves. Surely a chess master, a player of great games and great matches, at bottom scorns a problemist's purely mathematical art. He has much of it in reserve himself, and can produce it in an emergency: 'if he had made such and such a move, then I had such and such a winning combination in mind.' But the 'great game' of chess is primarily psychological, a conflict between one trained intelligence and another, and not a mere collection of small mathematical theorems.

---

[*] I believe that it is now regarded as a *merit* in a problem that there should be many variations of the same type.

枯燥的形式之一。数学证明应该像一个明确的星座,而不是银河系里某个分散的星团。

国际象棋难题既要出人意料,又要一定程度的简练,出奇制胜地落子是必需的,棋盘上的每个棋子都要发挥它们自己的用处。但这种审美效果是累积的。同样必须要有的是(除非问题过于简单,无法做到真正的有趣),妙招之后应该还有非常多的变化,而每种变化又都需要能见招拆招。"如果兵走 B5,那么马走 R6[*];如果……那么……;如果……那么……"——倘若没有许多不同的应对,那么就失去了审美效果。所有这些都是真正的数学,它们都有自身的优点;但这种"通过枚举法的证明"(那些枚举的事例没有本质区别[**]),恰恰是真正的数学家们常常轻视的。

我倾向于认为,可以通过棋手们自身的感受来强化我的论点。当然,国际象棋大师,即伟大的运动赛事选手,在本质上是蔑视解题者纯粹运用数学技艺的。他胸有成竹,紧要关头应对自如:"如果他这样或者那样下棋,那么我也会有对应的胜招。"但国际象棋的"大博弈"主要是心理层面的,它是两个训练有素的头脑之间的较量,而不仅仅是比拼一些简单数学定理的集合。

---

[*] 国际象棋中并没有 R6 位置,疑原文有误。——编者注
[**] 我相信,如今在那些应该有许多同类型变体的问题里,这也被认为是有价值的。

19

I must return to my Oxford apology, and examine a little more carefully some of the points which I postponed in §6. It will be obvious by now that I am interested in mathematics only as a creative art. But there are other questions to be considered, and in particular that of the 'utility' (or uselessness) of mathematics, about which there is much confusion of thought. We must also consider whether mathematics is really quite so 'harmless' as I took for granted in my Oxford lecture.

A science or an art may be said to be 'useful' if its development increases, even indirectly, the material well-being and comfort of men, if it promotes happiness, using that word in a crude and commonplace way. Thus medicine and physiology are useful because they relieve suffering, and engineering is useful because it helps us to build houses and bridges, and so to raise the standard of life (engineering, of course, does harm as well, but that is not the question at the moment). Now some mathematics is certainly useful in this way; the engineers could not do their job without a fair working knowledge of mathematics, and mathematics is beginning to find applications even in physiology. So here we have a possible ground for a defence of mathematics; it may not be the best, or even a particularly strong defence, but it is one which we must examine. The 'nobler' uses of mathematics, if such they be, the uses which it shares with all creative art, will be irrelevant to our examination. Mathematics may, like poetry or music, 'promote and sustain a lofty habit of mind', and so increase the

19

我必须回顾在牛津大学的那份辩白,更仔细地审视我在第6节最后提到的一些观点。现今很明显,我只对把数学当作一种具有创造性的艺术感兴趣。但我还需要考虑一些其他问题,特别是数学的"功用"(或无用性)问题,人们对这个问题的困惑很多。我们还必须考量数学是否真的像我在牛津大学的演讲里说的那样,可以想当然地认为它是那么地"无害"。

如果一门科学或一种艺术能够促进人类的物质财富和生活舒适度(即使是间接地),提升人们的幸福感,那么就可以说它是"有用的"。因此,医学和生理学是有用的,因为它们减轻人们的痛苦;工程学是有用的,因为它帮助我们建造房屋和桥梁,提高生活水平(当然工程学也是有害的,但在这里不做讨论)。如今,有些数学确实在这方面很有用,不学好数学知识,工程师们就无法完成他们的工作,数学甚至还可以应用于生理学领域。现在,我们似乎有了一个为数学进行辩白的理由,它或许不是最好的,甚至可能没有那么充分,但确实值得讨论。如果说数学还可能存在某种"更高尚"的用途,就像所有创造性艺术所具有的一样,这将与我们的讨论无关。诚然,就像诗歌和音乐,数学也有助于"促进和保持一种崇高的思

happiness of mathematicians and even of other people; but to defend it on that ground would be merely to elaborate what I have said already. What we have to consider now is the 'crude' utility of mathematics.

维习惯",使数学家甚至是其他人获得幸福感;但是,基于这个理由的辩护,仅仅是对我刚才说法的详尽阐述。我们现在要考虑的是数学的"原生"功用。

20

All this may seem very obvious, but even here there is often a good deal of confusion, since the most 'useful' subjects are quite commonly just those which it is most useless for most of us to learn. It is useful to have an adequate supply of physiologists and engineers; but physiology and engineering are not useful studies for ordinary men (though their study may of course be defended on other grounds). For my own part I have never once found myself in a position where such scientific knowledge as I possess, outside pure mathematics, has brought me the slightest advantage.

It is indeed rather astonishing how little practical value scientific knowledge has for ordinary men, how dull and commonplace such of it as has value is, and how its value seems almost to vary inversely to its reputed utility. It is useful to be tolerably quick at common arithmetic (and that, of course, is pure mathematics). It is useful to know a little French or German, a little history and geography, perhaps even a little economics. But a little chemistry, physics, or physiology has no value at all in ordinary life. We know that the gas will burn without knowing its constitution; when our cars break down we take them to a garage; when our stomach is out of order, we go to a doctor or a drugstore. We live either by rule of thumb or on other people's professional knowledge.

However, this is a side issue, a matter of pedagogy, interesting only to schoolmasters who have to advise parents clamouring for a 'useful'

20

所有这些似乎都很明显,但即便如此也经常会使人产生很多困惑,因为最"有用"的学科通常就是那些对大多数人来说最没用的。有足够多的生理学家和工程师是有好处的,但对普通人而言,生理学和工程学并不是有用的学问(尽管出于别的原因,人们也会为这些学问辩解)。就个人而言,我从未发现在纯数学以外所掌握的科学知识,给我带来过什么好处。

科学知识对普通人的实用价值如此之小,如此枯燥和平常,并且似乎还和它所宣称的用途几乎是背道而驰,确实令人相当吃惊。在普通算术方面,算得快是有用的(当然,这是纯数学)。懂几句法语或德语、一些历史和地理知识,甚至是经济学知识也都是有用的。但是,些许化学、物理学或生理学知识则在日常生活中毫无价值。不知道汽油的化学成分,并不妨碍我们知道它会燃烧;当汽车发生故障时,我们会把它们送到维修厂;当胃不舒服时,我们会去看医生或买点药。我们靠经验或别人的专业知识生活。

然而,这还是次要问题,它和教育有关,只有那些不得不建议父母为他们的孩子提供"有用的"教育的老师们才会对它感兴趣。

education for their sons. Of course we do not mean, when we say that physiology is useful, that most people ought to study physiology, but that the development of physiology by a handful of experts will increase the comfort of the majority. The questions which are important for us now are, how far mathematics can claim this sort of utility, what kinds of mathematics can make the strongest claims, and how far the intensive study of mathematics, as it is understood by mathematicians, can be justified on this ground alone.

在我们说生理学有用的时候,当然不是指大多数人应该去学习生理学,我们指的是屈指可数的生理学专家的研究会提高大多数人的舒适度。现在,对我们而言重要的问题是,数学能在多大程度上断言这种功用,又是什么样的数学才最担得起这个说法,以及仅凭这一点能在多大程度上证明数学家所理解的对数学进行的深入研究是正确的。

21

It will probably be plain by now to what conclusions I am coming; so I will state them at once dogmatically and then elaborate them a little. It is undeniable that a good deal of elementary mathematics—and I use the word 'elementary' in the sense in which professional mathematicians use it, in which it includes, for example, a fair working knowledge of the differential and integral calculus—has considerable practical utility. These parts of mathematics are, on the whole, rather dull; they are just the parts which have least aesthetic value. The 'real' mathematics of the 'real' mathematicians, the mathematics of Fermat and Euler and Gauss and Abel and Riemann, is almost wholly 'useless' (and this is as true of 'applied' as of 'pure' mathematics). It is not possible to justify the life of any genuine professional mathematician on the ground of the 'utility' of his work.

But here I must deal with a misconception. It is sometimes suggested that pure mathematicians glory in the uselessness of their work*, and make it a boast that it has no practical applications. The imputation is usually based on an incautious saying attributed to Gauss, to the effect that, if mathematics is the queen of the sciences, then the theory of numbers is, because of its supreme uselessness, the queen of mathematics—I have

---

* I have been accused of taking this view myself. I once said that 'a science is said to be useful if its development tends to accentuate the existing inequalities in the distribution of wealth, or more directly promotes the destruction of human life', and this sentence, written in 1915, has been quoted (for or against me) several times. It was of course a conscious rhetorical flourish, though one perhaps excusable at the time when it was written.

21

到目前为止,我的结论可能已经很明确了,因此,我会马上直陈其是,然后再稍作解释。不可否认,大量的基础数学——这里的"基础"是以数学家的视角看待的,其中包括相当有用的微积分知识——是有相当大的实用价值的。总体而言,这部分数学内容相当枯燥,审美价值也是最小的。"真正"数学家的"真正"数学,即费马、欧拉[73]、高斯、阿贝尔和黎曼所研究的数学,几乎是完全"无用的"(无论是"应用"数学还是"纯"数学都是如此)。对于任何真正的职业数学家,都不可能仅凭他工作的"实用性"来评判其一生。

但在这里我必须澄清一个误解。有的时候,人们认为纯数学家以他们的工作毫无用处而自豪*,并夸耀它没有实际应用。这种说法通常是基于高斯说过的一句不太谨慎的名言,其大意是:如果数学是科学的皇后,那么数论就是数学的皇后,因为它毫无用处——我

---

* 有人指责我就持这种观点。我在 1915 年曾经写过:"如果一门科学的发展会使现有的财富分配不平等加剧,或是更直接地助长对人类生活的破坏,那么这门科学就是有用的。"而这句话曾多次被用来赞成或反对我。这当然是一种夸张的修辞,尽管在写这句话的时候,这种夸张或许是可以理解的。

never been able to find an exact quotation. I am sure that Gauss's saying (if indeed it be his) has been rather crudely misinterpreted. If the theory of numbers could be employed for any practical and obviously honourable purpose, if it could be turned directly to the furtherance of human happiness or the relief of human suffering, as physiology and even chemistry can, then surely neither Gauss nor any other mathematician would have been so foolish as to decry or regret such applications. But science works for evil as well as for good (and particularly, of course, in time of war); and both Gauss and lesser mathematicians may be justified in rejoicing that there is one science at any rate, and that their own, whose very remoteness from ordinary human activities should keep it gentle and clean.

从来没有找到过这句话确切的出处。我确信高斯的这句话（如果确实是他说的）被相当粗暴地曲解了。倘若数论能用于实用且明显高尚的目的，可以直接提升人类的幸福或者减轻他们的痛苦——就像生理学或化学那样，那么高斯和其他数学家肯定不会愚蠢到诋毁或遗漏这样的应用。但是，科学既为善也为恶（尤其是在战争时期）；无论如何，高斯和那些不及于他的数学家们都有理由为存在这样一门科学而高兴，也就是他们从事的数学，由于和人类日常活动没什么关系，因而保持了其纯良的面貌。

22

There is another misconception against which we must guard. It is quite natural to suppose that there is a great difference in utility between 'pure' and 'applied' mathematics. This is a delusion: there is a sharp distinction between the two kinds of mathematics, which I will explain in a moment, but it hardly affects their utility.

How do pure and applied mathematics differ from one another? This is a question which can be answered definitely and about which there is general agreement among mathematicians. There will be nothing in the least unorthodox about my answer, but it needs a little preface.

My next two sections will have a mildly philosophical flavour. The philosophy will not cut deep, or be in any way vital to my main theses; but I shall use words which are used very frequently with definite philosophical implications, and a reader might well become confused if I did not explain how I shall use them.

I have often used the adjective 'real', and as we use it commonly in conversation. I have spoken of 'real mathematics' and 'real mathematicians', as I might have spoken of 'real poetry' or 'real poets', and I shall continue to do so. But I shall also use the word 'reality', and with two different connotations.

In the first place, I shall speak of 'physical reality', and here again I shall be using the word in the ordinary sense. By physical reality I mean the material world, the world of day and night, earthquakes and eclipses,

22

我们必须警惕另外一种误解。人们很自然地会认为,"纯"数学和"应用"数学在功用方面的差别很大。这是一种错觉。虽然在这两种数学之间存在着巨大差异——稍后我会解释,但那几乎不会影响它们的功用。

纯数学和应用数学之间有什么不同呢?这是一个可以明确回答的问题,答案也是数学家们普遍认可的。我的回答一点也不离经叛道,但在回答之前需要略加铺垫。

下面两节会略带哲学味。这种哲学并不太深奥,也不怎么会影响到我的主要观点。不过,我将用一些具有明确哲学意义的词汇,如果我不解释它们的使用规则,读者很可能会感到困惑。

我经常使用形容词"真正的",就像人们在谈话中经常用到的那样。我说过"真正的数学"和"真正的数学家",正如我说过"真正的诗歌"和"真正的诗人"一样,我会继续这样讲。但我也将使用"实在"一词,它有两种不同的含义。

首先,我将使用"物理实在"这个词,这里只用了它的普通含义。我所说的物理实在指的是物质世界,是拥有白天和黑夜、地震

the world which physical science tries to describe.

I hardly suppose that, up to this point, any reader is likely to find trouble with my language, but now I am near to more difficult ground. For me, and I suppose for most mathematicians, there is another reality, which I will call 'mathematical reality'; and there is no sort of agreement about the nature of mathematical reality among either mathematicians or philosophers. Some hold that it is 'mental' and that in some sense we construct it, others that it is outside and independent of us. A man who could give a convincing account of mathematical reality would have solved very many of the most difficult problems of metaphysics. If he could include physical reality in his account, he would have solved them all.

I should not wish to argue any of these questions here even if I were competent to do so, but I will state my own position dogmatically in order to avoid minor misapprehensions. I believe that mathematical reality lies outside us, that our function is to discover or *observe* it, and that the theorems which we prove, and which we describe grandiloquently as our 'creations', are simply our notes of our observations. This view has been held, in one form or another, by many philosophers of high reputation from Plato onwards, and I shall use the language which is natural to a man who holds it. A reader who does not like the philosophy can alter the language: it will make very little difference to my conclusions.

和日食的世界，它是物理学试图描述的世界。

在这一点上，我认为读者不会发现我在用词上的困扰，但我确实遇到了一个大麻烦。对我，或者对大多数数学家而言，还存在另外一种实在，我称之为"数学实在"。无论是数学家还是哲学家，对数学实在的本质都没有达成共识。有些人认为它是"精神上的"，在某种意义上它是我们构造的；另一些人则认为它是客观的，是独立于我们之外的。倘若有人能对数学实在做出令人信服的解释，那么就能解决许多最难的形而上学的问题。如果他能把物理实在囊括到他的解释里，那么他就能解决所有关于形而上学的问题。

即使我有能力探讨这些问题，我也不愿在这里讨论。不过，我会直接说出自己的观点，以避免产生一些不太要紧的误解。我相信，数学实在存在于我们之外，我们要做的是发现和**观察**它，那些我们所证明的定理，并且夸夸其谈地把它们说成是我们"创造"的定理，只不过是我们的观察记录。自柏拉图以来，许多享有盛誉的哲学家都以这样或那样的方式表达过这个观点。我使用的语言对持这种观点的人来说是很自然的。不喜欢哲学的读者可以换种说法，这对我的结论没什么影响。

23

The contrast between pure and applied mathematics stands out most clearly, perhaps, in geometry. There is the science of pure geometry[*], in which there are many geometries, projective geometry, Euclidean geometry, non-Euclidean geometry, and so forth. Each of these geometries is a *model*, a pattern of ideas, and is to be judged by the interest and beauty of its particular pattern. It is a *map* or *picture*, the joint product of many hands, a partial and imperfect copy (yet exact so far as it extends) of a section of mathematical reality. But the point which is important to us now is this, that there is one thing at any rate of which pure geometries are *not* pictures, and that is the spatio-temporal reality of the physical world. It is obvious, surely, that they cannot be, since earthquakes and eclipses are not mathematical concepts.

This may sound a little paradoxical to an outsider, but it is a truism to a geometer; and I may perhaps be able to make it clearer by an illustration. Let us suppose that I am giving a lecture on some system of geometry, such as ordinary Euclidean geometry, and that I draw figures on the blackboard to stimulate the imagination of my audience, rough drawings of straight lines or circles or ellipses. It is plain, first, that the truth of the theorems which I prove is in no way affected by the quality of my drawings. Their function is merely to bring home my meaning to my hearers, and, if I can do that, there would be no gain in having

---

[*] We must of course, for the purpose of this discussion, count as pure geometry what mathematicians call 'analytical' geometry.

23

也许在几何学里,纯数学和应用数学之间的差别是最明显的。纯几何科学*是存在的,它包括了好几种几何——射影几何、欧氏几何、非欧几何,等等。这些几何都是一种**模型**、一种思想的模式,它们的评判依据是它所描述的特定模式的趣味性和美感。纯几何是一种**映像**或者是一幅**图像**,是很多人共同努力的结果,是数学实在的一部分中一份不完整的拷贝(不过就其自身而言又是完整的)。但这里有一点很重要,无论从什么角度看,纯几何并**不能**描写物理世界的时空实在。显然,它们不可能描写物理实在,因为地震和日食不是数学概念。

对外行来说,这可能听起来有点矛盾,但对几何学家而言,这是众所周知的,也许我可以用一个例子把它解释清楚。假设我在做一个关于几何系统的讲座,比方说在讲普通的欧氏几何时,我会在黑板上画一些图形,它们可以是直线、圆或椭圆的示意图,有助听众的直观想象。很明显,首先,我画的图不会影响我所证明的定理的正确性。这些图形的作用只不过是让听众们理解我的意思,如果

---

* 当然,为了讨论,我们必须把纯几何理解为数学家们所说的"分析"几何。

them redrawn by the most skilful draughtsman. They are pedagogical illustrations, not part of the real subject-matter of the lecture.

Now let us go a stage further. The room in which I am lecturing is part of the physical world, and has itself a certain pattern. The study of that pattern, and of the general pattern of physical reality, is a science in itself, which we may call 'physical geometry'. Suppose now that a violent dynamo, or a massive gravitating body, is introduced into the room. Then the physicists tell us that the geometry of the room is changed, its whole physical pattern slightly but definitely distorted. Do the theorems which I have proved become false? Surely it would be nonsense to suppose that the proofs of them which I have given are affected in any way. It would be like supposing that a play of Shakespeare is changed when a reader spills his tea over a page. The play is independent of the pages on which it is printed, and 'pure geometries' are independent of lecture rooms, or of any other detail of the physical world.

This is the point of view of a pure mathematician. Applied mathematicians, mathematical physicists, naturally take a different view, since they are preoccupied with the physical world itself, which also has its structure or pattern. We cannot describe this pattern exactly, as we can that of a pure geometry, but we can say something significant about it. We can describe, sometimes fairly accurately, sometimes very roughly, the relations which hold between some of its constituents, and compare them with the exact relations holding between constituents of some system of pure geometry. We may be able to trace a certain resemblance between the two sets of relations, and then the pure geometry will become interesting to physicists; it will give us, to that extent, a map which 'fits the facts' of the physical world. The geometer offers to the physicist a whole set of maps from which to choose. One map, perhaps, will fit the facts better

我能做到这一点，那么让最熟练的制图员重画这些图形并不会带来什么好处。那些图形只是教学插图，讲座的真正主题并不包括它们。

现在让我们更进一步。我讲课的房间是物理世界的一部分，它本身遵循自己的模式。研究它的模式或是物理实在的一般模式就是一门科学，我们称为"物理几何学"。假设现在房间里多了一个强力发电机，或者一个巨大的重物。然后物理学家告诉我们，这个房间的几何形状发生了改变，整个物理模式发生的扭曲虽然微不足道，但这种扭曲绝对是存在的。我所证明的定理会不成立吗？如果认为我的证明受到了某种影响，那显然是胡说八道，这就好比在说某位读者把茶水洒在印有莎士比亚戏剧的纸上会让那部作品发生变化一样。戏剧和印刷它的书页无关，"纯几何"和教室或者任何其他物理世界的方方面面也无关。

这是纯数学家的观点。应用数学家、数学物理学家，自然会持不同的观点，因为他们沉浸于物理世界本身，它有其自身的结构或模式。我们无法像描述纯几何那样准确地描述这种模式，但我们可以讨论关于它的有意义的事情。可以时而精确、时而粗略地描述它的某些组成部分之间的关系，并拿它们和纯几何体系里的一些组成部分之间的确切关系做比较。我们也许能在这两组关系之间找到某种相似之处，于是，物理学家就会对纯几何产生了兴趣。在某种程度上，它为我们提供了一种"符合物理世界事实"的映像。几何学家为物理学家提供了一整套可供选择的映像。某种映像或许会比其

than others, and then the geometry which provides that particular map will be the geometry most important for applied mathematics. I may add that even a pure mathematician may find his appreciation of this geometry quickened, since there is no mathematician so pure that he feels no interest at all in the physical world; but, in so far as he succumbs to this temptation, he will be abandoning his purely mathematical position.

他映像更符合事实，然后，这种提供特定映像的几何学就成了对应用数学而言非常重要的一种几何。我还要补充一点，即使是纯数学家，也会发现自己更加欣赏几何学，因为任何数学家都不会纯粹到对物理世界一点兴趣也没有。但是，一旦被这种诱惑吸引，他就会放弃纯数学的立场。

24

There is another remark which suggests itself here and which physicists may find paradoxical, though the paradox will probably seem a good deal less than it did eighteen years ago. I will express it in much the same words which I used in 1922 in an address to Section A of the British Association. My audience then was composed almost entirely of physicists, and I may have spoken a little provocatively on that account; but I would still stand by the substance of what I said.

I began by saying that there is probably less difference between the positions of a mathematician and of a physicist than is generally supposed, and that the most important seems to me to be this, that the mathematician is in much more direct contact with reality. This may seem a paradox, since it is the physicist who deals with the subject-matter usually described as 'real'; but a very little reflection is enough to show that the physicist's reality, whatever it may be, has few or none of the attributes which common sense ascribes instinctively to reality. A chair may be a collection of whirling electrons, or an idea in the mind of God: each of these accounts of it may have its merits, but neither conforms at all closely to the suggestions of common sense.

I went on to say that neither physicists nor philosophers have ever given any convincing account of what 'physical reality' is, or of how the physicist passes, from the confused mass of fact or sensation with which he starts, to the construction of the objects which he calls 'real'. Thus we

24

这里还有一种说法,物理学家可能认为它是个悖论,尽管这种悖论比起 18 年前可能已经少了许多。我会用与 1922 年向协会[74]A 组演讲时几乎相同的词汇来表达我的观点。那次的听众几乎全是物理学家,因此我的话可能略带挑衅,但是我仍然坚持自己说过的话。

我一开始就指出,数学家和物理学家的立场差别可能比人们通常认为的小。在我看来,最重要的是,数学家与真实的联系要直接得多。这似乎是一个悖论,因为研究通常被描述为"真实"事物的那些人是物理学家。但是,只要稍作思考就能知道,物理学家所谓的"实在"——不管它是什么——很少或根本没有在通常意义下描绘真实的天然属性。一把椅子可能是旋转的电子的集合,也可能是上帝心中的概念:对它的每一种描述都有其优点,但又都不完全符合通常意义下的真实情况。

我接着要说的是,物理学家和哲学家没有对什么是"物理实在"给出过令人满意的解释,而物理学家也说不出该如何将大量混乱的真实和感受转变为所谓的"真实"结构。因此,我们并不知道物理

cannot be said to know what the subject-matter of physics is; but this need not prevent us from understanding roughly what a physicist is trying to do. It is plain that he is trying to correlate the incoherent body of crude fact confronting him with some definite and orderly scheme of abstract relations, the kind of scheme which he can borrow only from mathematics.

A mathematician, on the other hand, is working with his own mathematical reality. Of this reality, as I explained in §22, I take a 'realistic' and not an 'idealistic' view. At any rate (and this was my main point) this realistic view is much more plausible of mathematical than of physical reality, because mathematical objects are so much more what they seem. A chair or a star is not in the least like what it seems to be; the more we think of it, the fuzzier its outlines become in the haze of sensation which surrounds it; but '2' or '317' has nothing to do with sensation, and its properties stand out the more clearly the more closely we scrutinize it. It may be that modern physics fits best into some framework of idealistic philosophy—I do not believe it, but there are eminent physicists who say so. Pure mathematics, on the other hand, seems to me a rock on which all idealism founders: 317 is a prime, not because we think so, or because our minds are shaped in one way rather than another, but *because it is so*, because mathematical reality is built that way.

学的主旨究竟是什么，然而，这并不妨碍我们大致了解物理学家想要做什么。显然，物理学家试图把他所面对的那些毫无条理的原始事实与某种明确且有序的抽象关系联系起来，而他只能从数学借鉴这种模式。

另外，数学家也在研究数学实在。对于这种实在，正如我在第22节里所解释的，我持"实在论"的观点，而不认为它是"唯心主义的"。无论如何，这种实在论的观点在数学上比在物理学上更可信——这也是我的主要观点，因为数学对象更接近于它们看起来的样子。椅子或恒星丝毫不像它们看上去的样子，我们越研究，它们就越来越被感觉的迷雾所笼罩，轮廓也越发模糊。但是"2"或"317"和感觉是无关的，我们研究得越仔细，它们的性质就越明确。也许现代物理学很符合唯心主义哲学的某些框架——虽然我并不相信，但有一些杰出的物理学家是支持这个观点的。在我看来，纯数学是所有具有理想主义的创始人的基础：317是素数，它既不是我们想出来的，也不是因为我们用了这种或那种思考方式，它是素数只是因为它本来就是，因为数学实在就是以这种方式构造的。

25

These distinctions between pure and applied mathematics are important in themselves, but they have very little bearing on our discussion of the 'usefulness' of mathematics. I spoke in §21 of the 'real' mathematics of Fermat and other great mathematicians, the mathematics which has permanent aesthetic value, as for example the best Greek mathematics has, the mathematics which is eternal because the best of it may, like the best literature, continue to cause intense emotional satisfaction to thousands of people after thousands of years. These men were all primarily pure mathematicians (though the distinction was naturally a good deal less sharp in their days than it is now); but I was not thinking only of pure mathematics. I count Maxwell and Einstein, Eddington and Dirac, among 'real' mathematicians. The great modern achievements of applied mathematics have been in relativity and quantum mechanics, and these subjects are, at present at any rate, almost as 'useless' as the theory of numbers. It is the dull and elementary parts of applied mathematics, as it is the dull and elementary parts of pure mathematics, that work for good or ill. Time may change all this. No one foresaw the applications of matrices and groups and other purely mathematical theories to modern physics, and it may be that some of the 'highbrow' applied mathematics will become 'useful' in as unexpected a way; but the evidence so far points to the conclusion that, in one subject as in the other, it is what is commonplace and dull that counts for practical life.

25

纯数学和应用数学之间的这些区别很重要,但它们对我们讨论数学的"有用性"几乎没什么影响。我在第 21 节提到的费马和其他伟大数学家们的"真正的"数学,都具有永恒的审美价值,例如最优秀的古希腊数学就有这种价值。数学是永恒的,因为最好的数学就像最优秀的文学作品,能在几千年后继续让成千上万的人在情感上得到满足。这些人主要是纯数学家(尽管在他们所处的时代,这种差别毫无疑问比如今小得多);但我考虑的不只是纯数学。我把麦克斯韦[75]、爱因斯坦、爱丁顿[76]和狄拉克[77]也算作"真正的"数学家。现代应用数学在相对论和量子力学方面取得了巨大成就,而这些学科目前几乎和数论一样"毫无用处"。它是那种枯燥而又基础的应用数学,就像纯数学里枯燥而又基础的部分一样,无论好坏都能用。时间或许会改变一切。没有人预见到矩阵、群和其他纯数学理论在现代物理学中的应用,或许某些"高深"的应用数学也会以某种意想不到的方式变得"有用"。但迄今为止,无论什么学科,对实际生活起到重要作用的都是那些司空见惯而又枯燥乏味的东西。

I can remember Eddington giving a happy example of the unattractiveness of 'useful' science. The British Association held a meeting in Leeds, and it was thought that the members might like to hear something of the applications of science to the 'heavy woollen' industry. But the lectures and demonstrations arranged for this purpose were rather a fiasco. It appeared that the members (whether citizens of Leeds or not) wanted to be entertained, and that 'heavy wool' is not at all an entertaining subject. So the attendance at these lectures was very disappointing; but those who lectured on the excavations at Knossos, or on relativity, or on the theory of prime numbers, were delighted by the audiences that they drew.

我记得爱丁顿曾举过一个很好的例子，说明了"有用的"科学并没有吸引力。英国科学促进会曾在利兹举办过一次会议，组织者认为与会者可能想了解科学在"厚毛纺织"行业中的应用。然而，为此安排的讲座和演示相当失败。似乎与会者（无论他们是不是利兹市民）都想放松一下，而"厚毛纺织"根本就不是一个有趣的话题。所以，去听这些讲座的人都非常失望。但那些讲授克诺索斯[78]考古发掘、相对论、素数理论的人都很开心，因为他们能吸引听众。

26

What parts of mathematics are useful?

First, the bulk of school mathematics, arithmetic, elementary algebra, elementary Euclidean geometry, elementary differential and integral calculus. We must except a certain amount of what is taught to 'specialist', such as projective geometry. In applied mathematics, the elements of mechanics (electricity, as taught in schools, must be classified as physics).

Next, a fair proportion of university mathematics is also useful, that part of it which is really a development of school mathematics with a more finished technique, and a certain amount of the more physical subjects such as electricity and hydromechanics. We must also remember that a reserve of knowledge is always an advantage, and that the most practical of mathematicians may be seriously handicapped if his knowledge is the bare minimum which is essential to him; and for this reason we must add a little under every heading. But our general conclusion must be that such mathematics is useful as is wanted by a superior engineer or a moderate physicist; and that is roughly the same thing as to say, such mathematics as has no particular aesthetic merit. Euclidean geometry, for example, is useful in so far as it is dull—we do not want the axiomatics of parallels, or the theory of proportion, or the construction of the regular pentagon.

26

哪部分数学才是有用的呢？

首先，中小学校里的大部分数学，如算术、初等代数、初等欧氏几何、初等微积分，都是有用的。我们必须排除一些"专业"知识，比如射影几何。在应用数学领域，力学原理是有用的（学校里教的电学归入物理学）。

其次，相当比例的大学数学也是有用的：一些是中小学数学的进阶，它们具备更完善的技巧；还有一些更像物理学的学科，如电学和流体力学。我们还必须牢记，知识储备永远是有益的，如果某些注重实用的数学家只掌握了他该掌握的知识的下限，那么他们就可能会存在严重缺陷；因此，我们在各方面都必须多懂一些。但我们的结论一定会是，这样的数学是有用的，因为它们是高级工程师或普通物理学家们所需要的。这大致相当于说，它们没有特别的美学价值。例如，那些枯燥的欧氏几何是有用的——我们不需要平行公理，也无须比例理论或构造正五边形。

One rather curious conclusion emerges, that pure mathematics is one the whole distinctly more useful than applied. A pure mathematician seems to have the advantage on the practical as well as on the aesthetic side. For what is useful above all is *technique*, and mathematical technique is taught mainly through pure mathematics.

I hope that I need not say that I am not trying to decry mathematical physics, a splendid subject with tremendous problems where the finest imaginations have run riot. But is not the position of an ordinary applied mathematician in some ways a little pathetic? If he wants to be useful, he must work in a humdrum way, and he cannot give full play to his fancy even when he wishes to rise to the heights. 'Imaginary' universes are so much more beautiful than this stupidly constructed 'real' one; and most of the finest products of an applied mathematician's fancy must be rejected, as soon as they have been created, for the brutal but sufficient reason that they do not fit the facts.

The general conclusion, surely, stands out plainly enough. If useful knowledge is, as we agreed provisionally to say, knowledge which is likely, now or in the comparatively near future, to contribute to the material comfort of mankind, so that mere intellectual satisfaction is irrelevant, then the great bulk of higher mathematics is useless. Modern geometry and algebra, the theory of numbers, the theory of aggregates and functions, relativity, quantum mechanics—no one of them stands the test much better than another, and there is no real mathematician whose life can be justified on this ground. If this be the test, then Abel, Riemann, and Poincaré wasted their lives; their contribution to human comfort was negligible, and the world would have been as happy a place without them.

于是出现了一个相当奇怪的结论,纯数学无疑在总体上比应用数学更有用。纯数学家似乎在实用性和美学方面都占优。因为最有用的是**技巧**,而数学技巧主要是由纯数学教授的。

我希望没被误解成这是在责难数学物理,它是一门极好的学科,充满了大量极富想象力的问题。但是,作为一个普通的应用数学家,会不会略感可悲呢?如果他想成为有用的人,那就必须从事单调的工作,即使他想更上一层楼,也不能充分发挥他的想象力。"想象的"宇宙比拙劣构造出的"真实"宇宙要美丽得多;而且,应用数学家依靠想象得到的多数最完美的成果,一旦创造出来就被否定了,因为它们与事实不符,这个理由虽然残酷,却足够充分。

显然,总体结论已然非常清楚。如果有用的知识——就像我们暂时一致认同的,是指那些现在或在不久的将来可能会对人类的物质生活做出贡献的知识,那么仅就它们是否可以让人们得到知识满足而言,是无关紧要的,于是,高等数学的绝大部分是无用的。现代几何、代数、数论、集合与函数理论、相对论、量子力学都经不起考验,也没有一个真正的数学家能在这种评价体系下证明自己的工作是有价值的。倘若这就是衡量标准,那么阿贝尔、黎曼和庞加莱[79]都是碌碌无为之辈,他们对人类舒适度的贡献微不足道,如果没有他们,这个世界也照样会幸福。

27

It may be objected that my concept of 'utility' has been too narrow, that I have defined it in terms of 'happiness' or 'comfort' only, and have ignored the general 'social' effects of mathematics on which recent writers, with very different sympathies, have laid so much stress. Thus Whitehead (who has been a mathematician) speaks of 'the tremendous effect of mathematical knowledge on the lives of men, on their daily avocations, on the organization of society'; and Hogben (who is as unsympathetic to what I and other mathematicians call mathematics as Whitehead is sympathetic) says that 'without a knowledge of mathematics, the grammar of size and order, we cannot plan the rational society in which there will be leisure for all and poverty for none' (and much more to the same effect).

I cannot really believe that all this eloquence will do much to comfort mathematicians. The language of both writers is violently exaggerated, and both of them ignore very obvious distinctions. This is very natural in Hogben's case, since he is admittedly not a mathematician; he means by 'mathematics' the mathematics which he can understand, and which I have called 'school' mathematics. *This* mathematics has many uses, which I have admitted, which we can call 'social' if we please, and which Hogben enforces with many interesting appeals to the history of mathematical discovery. It is this which gives his book its merit, since it enables him to make plain, to many readers who never have been and never will be mathematicians, that there is more in mathematics than they

27

有人可能会反对我对"效用"的理解太狭隘了,我只用"幸福感"或"舒适度"定义,不去考虑数学的一般"社会"效应,然而,近来的作者们则一直在以不同的方式强调数学的社会效益。因此,数学家怀特海指出:"数学知识对人们的生活、日常爱好,以及社会组织影响巨大。"霍格本(他对我和其他数学家所说的数学并不感兴趣,而怀特海则相反)说:"如果没有数学知识,没有大小和秩序的法则,那么我们就无法规划一个可以让人们乐享生活而免于贫困的合理社会。"(还有一些其他类似意思的说法。)

实际上,我并不相信所有这些雄辩能给数学家带来什么安慰。那两位作者的言辞都被严重夸大了,他们都忽略了一些非常显著的不同。就霍格本而言这很自然,因为人们公认他不是数学家。他所说的"数学"是指那些他能理解的数学,我称之为"中小学"数学。我承认这种数学很有用,如果愿意,我们还可以把它称为"社会"数学,而霍格本又用了许多数学发现史上有趣的需求来强化它。正是这一点才让他的书有了价值,因为这使他能够向许多不是、也永远不会成为数学家的读者们阐明,数学要比他们想象的丰富得多。

thought. But he has hardly any understanding of 'real' mathematics (as any one who reads what he says about Pythagoras's theorem, or about Euclid and Einstein, can tell at once), and still less sympathy with it (as he spares no pains to show). 'Real' mathematics is to him merely an object of contemptuous pity.

It is not lack of understanding or of sympathy which is the trouble in Whitehead's case; but he forgets, in his enthusiasm, distinctions with which he is quite familiar. The mathematics which has this 'tremendous effect' on the 'daily avocations of men' and on 'the organization of society' is not the Whitehead but the Hogben mathematics. The mathematics which can be used 'for ordinary purposes by ordinary men' is negligible, and that which can be used by economists or sociologists hardly rises to 'scholarship standard'. The Whitehead mathematics may affect astronomy or physics profoundly, philosophy very appreciably— high thinking of one kind is always likely to affect high thinking of another—but it has extremely little effect on anything else. Its 'tremendous effects' have been, not on men generally, but on men like Whitehead himself.

但他几乎不懂任何"真正的"数学（所有读过他关于毕达哥拉斯定理或关于欧几里得和爱因斯坦的论述的人都能发现这点），也谈不上什么感同身受（这正是他不遗余力要表达的）。对他而言，"真正的"数学只不过是可鄙而又可怜的对象。

怀特海的问题则不是不懂数学或不能对它感同身受，而是他忘了，他热忱对待的数学和他所熟悉的数学是有区别的。对"人们的日常爱好"和"社会组织"有"巨大影响"的数学，是霍格本说的数学，而不是怀特海说的那种。能被"大众用于普通目的"的数学是不值一提的，而那些被经济学家和社会学家所使用的数学，也几乎无法上升到"学术标准"。怀特海的数学可能会对天文学和物理学产生深刻影响，会对哲学造成明显影响——一种高级思想总是会影响另一种高级思想——但它对其他事物的影响微乎其微。它只会对怀特海这样的人造成"巨大影响"，普罗大众并不会有什么感觉。

28

There are then two mathematics. There is the real mathematics of the real mathematicians, and there is what I will call the 'trivial' mathematics, for want of a better word. The trivial mathematics may be justified by arguments which would appeal to Hogben, or other writers of his school, but there is no such defence for the real mathematics, which must be justified as art if it can be justified at all. There is nothing in the least paradoxical or unusual in this view, which is that held commonly by mathematicians.

We have still one more question to consider. We have concluded that the trivial mathematics is, on the whole, useful, and that the real mathematics, on the whole, is not; that the trivial mathematics does, and the real mathematics does not, 'do good' in a certain sense; but we have still to ask whether either sort of mathematics does *harm*. It would be paradoxical to suggest that mathematics of any sort does much harm in time of peace, so that we are driven to the consideration of the effects of mathematics on war. It is very difficult to argue such questions at all dispassionately now, and I should have preferred to avoid them; but some sort of discussion seems inevitable. Fortunately, it need not be a long one.

There is one comforting conclusion which is easy for a real mathematician. Real mathematics has no effects on war. No one has yet discovered any warlike purpose to be served by the theory of numbers or relativity, and it seems very unlikely that anyone will do so for many

28

有两种数学。一种是真正的数学家研究的"真正的"数学,另一种则是我所谓的"平凡的"数学——没有更好的词来形容这种数学了。平凡的数学可以用霍格本或他那一派的作者提出的论据来辩护,而真正的数学却没有这种辩词,即使它能被辩护,也只能把它当作艺术。这个观点一点也不含混,也没什么特别,数学家们普遍都是这么想的。

还有一个问题需要考虑。我们刚刚已经得出了结论:总体而言,平凡的数学是有用的,而真正的数学是无用的。就某种意义而言,平凡的数学是有益的,而真正的数学并不是。但我们不禁要问,这两种数学是否是**有害的**呢?认为数学在和平时期会造成巨大危害是荒谬的,因此我们只考虑数学对战争的影响。现在要冷静地探讨这类问题是很困难的,我也应避免讨论这些问题。然而,这个讨论似乎又是无法回避的。幸好并不需要讨论很久。

对真正的数学家而言,很容易得到一个令人欣慰的结论:真正的数学对战争没什么用。迄今为止,还没有人发现数论或相对论能被用于战争目的,而且在未来的很长一段日子里,似乎也不太可能

years. It is true that there are branches of applied mathematics, such as ballistics and aerodynamics, which have been developed deliberately for war and demand a quite elaborate technique: it is perhaps hard to call them 'trivial', but none of them has any claim to rank as 'real'. They are indeed repulsively ugly and intolerably dull; even Littlewood could not make ballistics respectable, and if he could not who can? So a real mathematician has his conscience clear; there is nothing to be set against any value his work may have; mathematics is, as I said at Oxford, a 'harmless and innocent' occupation.

The trivial mathematics, on the other hand, has many applications in war. The gunnery experts and aeroplane designers, for example, could not do their work without it. And the general effect of these applications is plain: mathematics facilitates (if not so obviously as physics or chemistry) modern, scientific, 'total' war.

It is not so clear as it might seem that this is to be regretted, since there are two sharply contrasted views about modern scientific war. The first and the most obvious is that the effect of science on war is merely to magnify its horror, both by increasing the sufferings of the minority who have to fight and by extending them to other classes. This is the most natural and the orthodox view. But there is a very different view which seems also quite tenable, and which has been stated with great force by Haldane in *Callinicus*[*]. It can be maintained that modern warfare is *less* horrible than the warfare of pre-scientific times; that bombs are probably more merciful than bayonets; that lachrymatory gas and mustard gas are

---

[*] J. B. S. Haldane, *Callinicus: a Defence of Chemical Warfare* (1924).

有人会这么做。诚然，应用数学的某些分支，如弹道学和空气动力学，是为战争而特意发展起来的，它们需要相当复杂的技术——也许很难把它们归为"平凡的"数学，但它们同样也都没有资格被当作"真正的"数学。它们确实丑陋得令人生厌，也枯燥得让人作呕，即使有李特尔伍德加盟，也无法让人们对弹道学产生敬意。如果他不能，又有谁可以做到呢？所以真正的数学家是问心无愧的，他们的工作可能具有的所有价值都是无可非议的。正如我在牛津大学所说，数学是一种"无害而清白"的职业。[80]

另外，平凡的数学在战争中有许多应用。例如，倘若没有这种数学，射击专家和飞机设计师就无法工作。这些应用的总体效果是明显的，数学促进了（即使不像物理学或化学那么明显）现代的、科学的、全面的战争。

关于现代科学化的战争有两种截然相反的观点，因此，它并不像看起来那么明显地令人遗憾。首先，也是最明显的一点，科学对战争的影响仅仅是放大战争的恐怖，这种恐怖是通过增加不得不参战的少数人的痛苦，并且将其不断扩展到其他阶层来实现的。这是最自然，也是最正统的观点。但还有一种非常不一样的说法，似乎也很有道理，霍尔丹[81]的《卡里尼科斯[82]》就极力主张这种观点*。这种观点认为，现代战争并没有前科学时代的战争那么可怕：炸弹可能比刺刀更仁慈，而通过军事科学发明的催泪瓦斯和芥子气也许

---

* J. B. S. 霍尔丹，《卡里尼科斯：化学战争的防御》（1924）。

perhaps the most humane weapons yet devised by military science; and that the orthodox view rests solely on loose-thinking sentimentalism[*]. It may also be urged (though this was not one of Haldane's theses) that the equalization of risks which science was expected to bring would be in the long run salutary; that a civilian's life is not worth more than a soldier's, nor a woman's than a man's; that anything is better than the concentration of savagery on one particular class; and that, in short, the sooner war comes 'all out' the better.

I do not know which of these views is nearer to the truth. It is an urgent and a moving question, but I need not argue it here. It concerns only the 'trivial' mathematics, which it would be Hogben's business to defend rather than mine. The case for his mathematics may be rather more than a little soiled; the case for mine is unaffected.

Indeed, there is more to be said, since there is one purpose at any rate which the real mathematics may serve in war. When the world is mad, a mathematician may find in mathematics an incomparable anodyne. For mathematics is, of all the arts and sciences, the most austere and the most remote, and a mathematician should be of all men the one who can most easily take refuge where, as Bertrand Russell says, 'one at least of our nobler impulses can best escape from the dreary exile of the actual world'. It is a pity that it should be necessary to make one very serious reservation—he must not be too old. Mathematics is not a contemplative but a creative subject; no one can draw much consolation from it when he has lost the power or the desire to create; and that is apt to happen to a mathematician rather soon. It is a pity, but in that case he does not matter a great deal anyhow, and it would be silly to bother about him.

---

[*] I do not wish to prejudge the question by this much misused word; it may be used quite legitimately to indicate certain types of unbalanced emotion. Many people, of course, use 'sentimentalism' as a term of abuse for other people's decent feelings, and 'realism' as a disguise for their own brutality.

是最人道的武器。正统的观点只不过是建立在思考不周的感伤主义[*]之上。它还鼓吹说（尽管这不是霍尔丹提出的），科学带来的风险均衡化从长远来看是有益的：平民的生命并不比士兵的更宝贵，女人的生命也不比男人的更有价值。任何情况都比把暴行集中在某个特定阶层更好。简言之，"全面铺开"的战争越早爆发越好。

我不知道这两种观点中的哪一种更接近事实。这是一个要紧并且不断变化的问题，但我不必在此争辩。它只涉及"平凡的"数学，捍卫这种数学与霍格本有关，而与我无关。他说的数学可能有不少污点，但我的数学并没有。

事实上，我还要补充几句，因为不管怎样，真正的数学可能在战争中还会用于这样的目的：当世界处于癫狂状态时，数学家或许能在数学里发现一种无与伦比的止痛剂。在所有艺术和科学里，数学是最朴实、最遁世的，而数学家也应该是所有人里最容易隐居的。伯特兰·罗素说："在我们高尚的冲动里，至少有一种能很好地让我们从寂苦游荡着的现实世界里解脱出来。"很可惜，这种说法还必须加上一个非常重要的条件——他一定不能太老。数学不是冥想术，它是一门创造性的学科。当一个人失去了创造的能力和欲望时，他并不能从中获取多少慰藉。这种事情很容易也很快就会在数学家身上发生。这是令人遗憾的，不过在这种情况下，他也不再那么重要，为他操心是愚蠢的。

---

[*] 我并不希望用这个非常不准确的字眼来预判该问题，它是可以用来非常合理地表示某种失衡情绪的。当然，有很多人会用"感伤主义"来错误地形容别人的正常感受，并且用"现实主义"来掩饰自己的残酷无情。

29

I will end with a summary of my conclusions, but putting them in a more personal way. I said at the beginning that anyone who defends his subject will find that he is defending himself; and my justification of the life of a professional mathematician is bound to be, at bottom, a justification of my own. Thus this concluding section will be in its substance a fragment of autobiography.

I cannot remember ever having wanted to be anything but a mathematician. I suppose that it was always clear that my specific abilities lay that way, and it never occurred to me to question the verdict of my elders. I do not remember having felt, as a boy, any *passion* for mathematics, and such notions as I may have had of the career of a mathematician were far from noble. I thought of mathematics in terms of examinations and scholarships: I wanted to beat other boys, and this seemed to be the way in which I could do so most decisively.

I was about fifteen when (in a rather odd way) my ambitions took a sharper turn. There is a book by 'Alan St Aubyn'* called *A Fellow of Trinity*, one of a series dealing with what is supposed to be Cambridge college life. I suppose that it is a worse book than most of Marie Corelli's; but a book can hardly be entirely bad if it fires a clever boy's imagination. There are

---

* 'Alan St Aubyn' was Mrs Frances Marshall, wife of Matthew Marshall.

29

最后,我用更个人的表达方式来总结一下。我在一开始就说,任何为自己从事的学科辩白的人,都会发现是在为自己说话。我对职业数学家生涯的辩白,本质上必然也是为我自己说话。因此,最后一节实质上是一段我的自传。

我只记得自己想当一名数学家。我一直很明确地认为,自己的特殊能力就是数学,我也从未想过询问长辈们的意见。我已经不记得在我还是个孩子的时候,对数学有过什么激情,我对数学家的职业生涯可能有过这样的想法,但这些想法远非高尚。我是从考试和奖学金的角度来衡量数学的:我想胜过其他男孩,而数学似乎是我能达到这个目的的最佳方式。

在大约 15 岁时,我的抱负(以一种相当奇怪的方式)发生了巨大变化。艾伦·圣·奥宾[*]曾写过一本名为《一位三一学院的研究员》的书,它是一系列关于剑桥大学生活的图书之一。我觉得这本书不如玛丽·科雷利[83]写的大多数作品。但是,如果一本书能激发一个聪明男孩的想象力,它就不是那么一无是处。书里提到了两位人物,

---

[*] "艾伦·圣·奥宾"就是弗朗西丝·马歇尔女士,她是马修·马歇尔的妻子。

two heroes, a primary hero called Flowers, who is almost wholly good, and a secondary hero, a much weaker vessel, called Brown. Flowers and Brown find many dangers in university life, but the worst is a gambling saloon in Chesterton[*] run by the Misses Bellenden, two fascinating but extremely wicked young ladies. Flowers survives all these troubles, is Second Wrangler and Senior Classic, and succeeds automatically to a Fellowship (as I suppose he would have done then). Brown succumbs, ruins his parents, takes to drink, is saved from delirium tremens during a thunderstorm only by the prayers of the Junior Dean, has much difficult in obtaining even an Ordinary Degree, and ultimately becomes a missionary. The friendship is not shattered by these unhappy events, and Flowers's thoughts stray to Brown, with affectionate pity, as he drinks port and eats walnuts for the first time in Senior Combination Room.

Now Flowers was a decent enough fellow (so far as 'Alan St Aubyn' could draw one), but even my unsophisticated mind refused to accept him as clever. If he could do these things, why not I? In particular, the final scene in Combination Room fascinated me completely, and from that time, until I obtained one, mathematics meant to me primarily a Fellowship of Trinity.

I found at once, when I came to Cambridge, that a Fellowship implied 'original work', but it was a long time before I formed any definite idea of research. I had of course found at school, as every future mathematician does, that I could often do things much better than my teachers; and even at Cambridge I found, though naturally much less

---

[*] Actually, Chesterton lacks picturesque features.

主角叫弗劳尔斯,他几乎是完美的,配角布朗则是一个意志薄弱的人。弗劳尔斯和布朗在大学生活中遇到了很多危险,其中最糟糕的是在切斯特顿*的一家赌场,它是由年轻迷人但又极其邪恶的贝伦登两姐妹经营的。弗劳尔斯战胜重重困难,取得了数学学位考试第二名以及古典文学学位考试第一名的佳绩,最后自动成了研究员(我想他当时一定是这样的)。然而,布朗被击溃了,他辜负了父母并开始酗酒,在一次暴风雨中,一位年轻的牧师通过祷告把他从酒后的神志不清中拯救了出来,他甚至连得到一个普通的学位都很困难,最后当了一名传教士。然而这些不幸并没有破坏他们的友谊,弗劳尔斯在高级学术交流室第一次品尝着波特酒和核桃时,满怀深切同情地想到了布朗。

(根据艾伦·圣·奥宾的描写)弗劳尔斯是一个相当正派的人,然而淳朴的我并不认为他很聪明。如果他都能做到这些,我为什么就不能呢?特别是,我完全着迷于他在高级学术交流室的最后一幕。从那时起一直到我成为学院研究员,数学对我来说就意味着三一学院的研究员资格。

我来到剑桥后立刻发现,研究员意味着"原创性工作",但直到很久以后,我才对研究工作有了明确的想法。当然,我在中小学时就知道,正如每一位未来的数学家那样,我常常能做得比老师好得多。即便是在剑桥,我也发现了这点,尽管不那么频繁,但有时

---

* 实际上,切斯特顿并没什么特别之处。

frequently, that I could sometimes do things better than the College lecturers. But I was really quite ignorant, even when I took the Tripos, of the subjects on which I have spent the rest of my life; and I still thought of mathematics as essentially a 'competitive' subject. My eyes were first opened by Professor Love, who taught me for a few terms and gave me my first serious conception of analysis. But the great debt which I owe to him—he was, after all, primarily an applied mathematician— was his advice to read Jordan's famous *Cours d'analyse*; and I shall never forget the astonishment with which I read that remarkable work, the first inspiration for so many mathematicians of my generation, and learnt for the first time as I read it what mathematics really meant. From that time onwards I was in my way a real mathematician, with sound mathematical ambitions and a genuine passion for mathematics.

I wrote a great deal during the next ten years, but very little of any importance; there are not more than four or five papers which I can still remember with some satisfaction. The real crises of my career came ten or twelve years later, in 1911, when I began my long collaboration with Littlewood, and in 1913, when I discovered Ramanujan. All my best work since then has been bound up with theirs, and it is obvious that my association with them was the decisive event of my life. I still say to myself when I am depressed, and find myself forced to listen to pompous and tiresome people, 'Well, I have done one thing *you* could never have done, and that is to have collaborated with both Littlewood and Ramanujan on something like equal terms.' It is to them that I owe an unusually late maturity: I was at my best at a little past forty, when I was a professor at Oxford. Since then I have suffered from that steady deterioration which is the common fate of elderly men and particularly of

我能比讲师们做得更好。然而，即便是在我通过优等考试时，我对余生将要研究的那些科目仍然一无所知。当时我认为数学本质上是一门"竞争性"学科。洛夫[84]教授让我第一次大开眼界，他教了我几个学期，让我第一次懂得了关于分析的严格概念。但我最要感谢他的，是他建议我读若尔当[85]著名的《分析教程》——说到底他主要还是个应用数学家。我永远也不会忘记，在我读到这本杰作时的惊叹之情。对我这一代的众多数学家而言，它是第一流的启智之作，我也从中第一次悟到了数学的真正含义。从那时起，我开始怀着雄心勃勃的数学抱负和真挚的数学热情，用自己的方式去成为一名真正的数学家。

在接下来的十年里，我做了很多研究，但几乎没有什么重要的成果，我至今还记得，能令自己满意的论文不会超过四到五篇。我真正的事业危机出现在十年或十二年之后。1911 年，我开始了与李特尔伍德的长期合作，1913 年，我发现了拉马努金。从那时起，我所有最好的作品都与他们有关，毫无疑问，与他们相识是我一生中的决定性事件。在我意志消沉、发现自己不得不听那些自负而又令人生厌的人呱噪的时候，我仍然会对自己说："好吧，我做了一件你们永远也不可能办到的事，那就是在平等的情况下与李特尔伍德和拉马努金合作。"对他们而言，我格外晚熟：当我成为牛津大学的教授时，正处于最佳状态，那时我已四十出头。此后，我便陷于不断退化的痛苦之中，这是年长之人，特别是年长的数学家的共同宿

elderly mathematicians. A mathematician may still be competent enough at sixty, but it is useless to expect him to have original ideas.

It is plain now that my life, for what it is worth, is finished, and that nothing I can do can perceptibly increase or diminish its value. It is very difficult to be dispassionate, but I count it a 'success'; I have had more reward and not less than was due to a man of my particular grade of ability. I have held a series of comfortable and 'dignified' positions. I have had very little trouble with the duller routine of universities. I hate 'teaching', and have had to do very little, such teaching as I have done having been almost entirely supervision of research; I love lecturing, and have lectured a great deal to extremely able classes; and I have always had plenty of leisure for the researches which have been the one great permanent happiness of my life. I have found it easy to work with others, and have collaborated on a large scale with two exceptional mathematicians; and this has enabled me to add to mathematics a good deal more than I could reasonable have expected. I have had my disappointments, like any other mathematician, but none of them has been too serious or has made me particularly unhappy. If I had been offered a life neither better nor worse when I was twenty, I would have accepted without hesitation.

It seems absurd to suppose that I could have 'done better'. I have no linguistic or artistic ability, and very little interest in experimental science. I might have been a tolerable philosopher, but not one of a very original kind. I think that I might have made a good lawyer; but journalism is the only profession, outside academic life, in which I should have felt really confident of my chances. There is no doubt that I was right to be a mathematician, if the criterion is to be what is commonly called success.

命。也许有人在六十岁后还能胜任数学家的工作，但已经不能指望他还会有什么独到的见解了。

很显然，我有价值的那部分生命如今已经结束，不管我做什么都不能明显地对其价值产生影响。要保持心平气和是很难的，但我把它当作一种"成功"。我得到的回报已经超出具有我这种级别能力的人所应得到的，而不是不足。我也曾担任过一系列舒适而"高贵"的职务。我不觉得大学里枯燥的日常生活有什么问题。我讨厌"教学"工作，并且尽可能少地参与其中，我所做的几乎全都是指导研究；我喜欢开讲座，并且也为很多能力超群的班级做过讲座；我总是有足够的闲暇时间做研究，这是我一生中最大的恒久幸福。我发现与他人合作很容易，并和两位杰出的数学家进行了广泛合作，这使我为数学贡献了许多意料之外的成果。和其他数学家一样，我也有过失望，但这些失望既没有很严重的，也没有让我特别不开心的。倘若在我二十岁时，有人给我一种比那时不好也不坏的生活，那么我会毫不犹豫地接受它。

认为我本可以"做得更好"似乎很荒谬。我没有语言和艺术的才能，对实验科学也没什么兴趣。我或许能成为一名还算过得去的哲学家，但不会是那种原创型的。我也有可能成为一个好律师，但除了学术生涯，新闻业或许才是我唯一应该有真正的信心改投的职业。如果标准是人们所谓的成功的话，那么毫无疑问成为一名数学家是我正确的选择。

My choice was right, then, if what I wanted was a reasonably comfortable and happy life. But solicitors and stockbrokers and bookmakers often lead comfortable and happy lives, and it is very difficult to see how the world is the richer for their existence. Is there any sense in which I can claim that my life has been less futile than theirs? It seems to me again that there is only one possible answer: yes, perhaps, but, if so, for one reason only.

I have never done anything 'useful'. No discovery of mine has made, or is likely to make, directly or indirectly, for good or ill, the least difference to the amenity of the world. I have helped to train other mathematicians, but mathematicians of the same kind as myself, and their work has been, so far at any rate as I have helped them to it, as useless as my own. Judged by all practical standards, the value of my mathematical life is nil; and outside mathematics it is trivial anyhow. I have just one chance of escaping a verdict of complete triviality, that I may be judged to have created something worth creating. And that I have created something is undeniable: the question is about its value.

The case for my life, then, or for that of any one else who has been a mathematician in the same sense in which I have been one, is this: that I have added something to knowledge, and helped others to add more; and that these somethings have a value which differs in degree only, and not in kind, from that of the creations of the great mathematicians, or of any of the other artists, great or small, who have left some kind of memorial behind them.

如果我想要的是一种相当舒适、幸福的生活，那么我的选择是正确的。律师、股票经纪人和出版商往往也过着舒适而幸福的生活，但很难看出世界会因他们的存在而变得更富有。从某种角度而言，我可以说我的生活比他们的更充实吗？在我看来，答案似乎只有一种：是的。也许的确如此，但如果真是这样，那么原因也只有一个。

我从未做过什么"有用的"事。我的发现都没有（并且看起来也不会）对这个世界的便利性产生什么影响，不管这种影响是直接的还是间接的，是积极的还是消极的。我帮助培养了一些数学家，但这些数学家也和我一样，至少到目前为止在我的指引下，他们的工作和我的一样没什么用。如果全部以"实用"的标准来衡量，那么我的数学生命的价值是零；而在数学之外，它更是普普通通。我只有一种机会可以摆脱被完全忽视，那就是让人们认为我创造了一些值得创造出来的东西。当然，我所创造的东西是不可否认的，问题在于它们有什么价值。

我的一生，或者任何一个和我情况相仿的数学家一生的价值，就是创造了一些新知识，也帮助别人创造了一些。相较于那些伟大数学家的创造，或者其他大大小小的艺术家们身后留下的各种纪念物，这些新知识的价值只是在程度上有所不同，但它们的性质是一样的。

Note

Professor Broad and Dr Snow have both remarked to me that, if I am to strike a fair balance between the good and evil done by science, I must not allow myself to be too much obsessed by its effects on war; and that, even when I am thinking of them, I must remember that it has many very important effects besides those which are purely destructive. Thus (to take the latter point first), I must remember (a) that the organization of an entire population for war is only possible through scientific methods; (b) that science has greatly increased the power of propaganda, which is used almost exclusively for evil; and (c) that it has made 'neutrality' almost impossible or unmeaning, so that there are no longer 'islands of peace' from which sanity and restoration might spread out gradually after war. All this, of course, tends to reinforce the case *against* science. On the other hand, even if we press this case to the utmost, it is hardly possible to maintain seriously that the evil done by science is not altogether outweighed by the good. For example, if ten million lives were lost in every war, the net effect of science would still have been to increase the average length of life. In short, my §28 is much too 'sentimental'.

I do not dispute the justice of these criticisms, but, for the reasons which I state in my preface, I have found it impossible to meet them in my text, and content myself with this acknowledgement.

Dr Snow has also made an interesting minor point about §8. Even if we grant that 'Archimedes will be remembered when Aeschylus is

注记

布罗德教授和斯诺博士都对我说过,如果我要在科学带来的善与恶之间主持公道,就不能让自己过度沉迷于科学对战争的影响;而且,即便在考虑它们时,我也必须牢记,除了那些纯粹的破坏,还有许多非常重要的影响。因此(首先基于后者),我必须记住:(a) 只有通过科学的方法才能组织起全员参加战争;(b) 科学大大增强了宣传的能力,而宣传几乎全都是用于作恶的;(c) 科学让"中立"变得几乎不可能或毫无意义,因此不再会有"世外桃源",能让人们在战争结束后可以从这些地方逐步恢复理智和秩序。当然,所有这些都是站在反对科学一方的。但另一方面,即使我们把这个问题推到极致,也不能真的认为科学做的恶一定大于科学行的善。例如,即便在每一场战争中都有 1000 万人丧生,科学的净效应仍使人类的平均寿命延长了。简言之,我写的第 28 节太过"感伤"了。

我并不想争辩这些批评的正义性,但出于我在前言中所述的原因,我发现要在正文里加以讨论,并且让自己承认这点是不可能的。

斯诺博士针对第 8 节还提出了一个有趣的观点。即使我们承认"当埃斯库罗斯被遗忘时,阿基米德仍将被铭记",难道数学上的名

forgotten', is not mathematical fame a little too 'anonymous' to be wholly satisfying? We could form a fairly coherent picture of the personality of Aeschylus (still more, of course, of Shakespeare or Tolstoi) from their works alone, while Archimedes and Eudoxus would remain mere names.

Mr J. M. Lomas put this point more picturesquely when we were passing the Nelson column in Trafalgar square. If I had a statue on a column in London, would I prefer the columns to be so high that the statue was invisible, or low enough for the features to be recognizable? I would choose the first alternative, Dr Snow, presumably, the second.

望如此之"低调"就能令人完全满意吗？我们可以仅仅通过埃斯库罗斯（当然，也包括莎士比亚或托尔斯泰[86]）的作品，就对他的性格形成一个相当清晰的印象，但阿基米德和欧多克索斯留下的只不过是名字。

当我们经过特拉法加广场[87]的纳尔逊[88]纪念柱时，约翰·洛马斯[89]先生把这一点表达得更加生动。倘若我在伦敦的纪念柱上有一座雕像，我是希望纪念柱高得看不见雕像，还是希望它低到能让人们辨识出雕像是谁呢？我会选择第一种，而斯诺博士也许会选第二种。

译后记

哈代的名著《一个数学家的辩白》已有多个译本。在人民邮电出版社图灵公司发出翻译邀请时,虽然我已经翻译了《数学万花筒3:夏尔摩斯探案集》[90]和《不可思议的数》[91],但还是有些胆怯的。好在有江志强老师的支持鼓励,才安心许多,在此向江老师表达深深的谢意。

这本小册子虽然名声在外,但其中的一些观点其实是值得商榷的。比如,哈代认为数学(以及文中提到的近代物理)不需要也没有功用性,如今的人们恐怕很难再认同这类观点。又比如数学和年龄问题,倘若哈代在世,他可能觉得阿蒂亚爵士[93]去年召开的黎曼猜想讲座就是在"炒作",但我们更愿意相信,只要有"雄心"让今天的自己胜于昨天,那么哪怕再小的贡献,都是"持久性"的,因为这是数学的特性。

本书最后一章提到的艾伦·圣·奥宾写书的那个小故事,让我颇为振奋。但愿众多数学爱好者们创作、改写、翻译、传播的"关于数学的"作品,能激发更多有数学天赋的人们爱上数学,不断地续写出"邂逅数学——缔造传奇"这样的动人故事。

最后，我为本书绘制了一幅各节关系简图以供参考。感谢陈见柯老师和我的家人朋友的帮助，也欢迎读者发送邮件到 dr.watsup@outlook.com，对译本提出批评和指正。希望经过若干次迭代后，本书能成为 A Mathematician's Apology 的译本佳作。

何生

2019 年 9 月

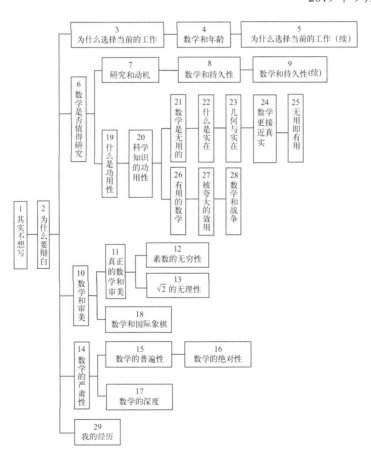

译者注

这里给出了文中提及人物或作品的简单介绍。

[1] John Millington Lomas（1917—1945），板球运动员，哈代的密友。

[2] Charlie Dunbar Broad（1887.12.30—1971.03.11），英国科学哲学家、作家。

[3] Charles Percy Snow（1905.10.15—1980.07.01），英国物理化学家、小说家，哈代的朋友。

[4] 指《战争时代的数学》(*Mathematics in war-time*)。

[5] Alfred Edward Housman（1859.03.26—1936.04.30），英国古典文化学者、诗人。

[6] Leslie Stephen（1832.11.28—1904.02.22），英国作家。

[7] 本文写于 1940 年，而豪斯曼在 1936 年就去世了。

[8] Albert Einstein（1879.03.14—1955.04.18），德国/瑞士/美国科学家、物理学家。

[9] Francis Herbert Bradley（1846.01.30—1924.09.18），英国哲学家。

[10] Samuel Johnson（1709.09.18—1784.12.03），英国作家、文学评论家和诗人。

[11] 本段引自《塞缪尔·约翰逊的生平》，因此这里的"我"并不是哈代。

[12] Alexander Alekhine（1892.10.31—1946.03.24），苏联国际象棋运动员。

[13] Don Bradman（1908.08.27—2001.02.25），澳大利亚板球运动员。

[14] Walter James Redfern Turner（1889.10.13—1946.11.18），英国作家、批评家。

[15] Victor Thomas Trumper（1877.11.02—1915.06.28），澳大利亚板球运动员。

[16] Rupert Chawner Brooke（1887.08.03—1915.04.23），英国诗人。

[17] Isaac Newton（1643.01.04—1727.03.31），英国数学家、物理学家。

[18] Évariste Galois（1811.10.25—1832.05.31），法国数学家。

[19] Niels Henrik Abel（1802.08.05—1829.04.06），挪威数学家，证明了高于四次的代数方程没有一般形式的代数解。

[20] Srinivasa Ramanujan（1887.12.22—1920.04.26），印度数学家。

[21] Georg Friedrich Bernhard Riemann（1826.09.17—1866.07.20），德国数学家。

[22] Johann Carl Friedrich Gauß（1777.04.30—1855.02.23），德国著名数学家、物理学家、天文学家、几何学家、大地测量学家。

[23] Paul Painlevé（1863.12.05—1933.10.29），法国数学家、政治家、航空理论家。

[24] Pierre-Simon Laplace (1749.03.23—1827.03.05)，法国数学家、物理学家、天文学家，他在60多岁时在概率论和数学物理方法方面做出了重大贡献。

[25] Blaise Pascal（1623.06.19—1662.08.19），法国数学家、物理学家、神学家，他在1654年皈依后，放弃数学转而研究神学和哲学。

[26] Pythagoras of Samos（约公元前570—约公元前495），古希腊宗教领袖、数学家。

[27] Attila the Hun（约406—453），古代亚欧大陆匈人的领袖和皇帝。

[28] Napoléon Bonaparte（1769.08.15—1821.05.05），法国军事家和政治家，法兰西第一帝国皇帝。

[29] Joseph Lister（1827.04.05—1912.02.10），英国外科医生，外科手术消毒技术的发明者和推广者。

[30] Louis Pasteur（1822.12.27—1895.09.28），法国微生物学家、化学家，微生物学的奠基人之一。

[31] King Camp Gillette（1855.01.05—1932.07.09），美国商人，吉列剃须护理品牌创始人。

[32] William Willett（1856.08.10—1915.03.04），英国建筑师，1907年向英国议会提出一种夏时制方案。

[33] Hammurabi（约公元前1810—约公元前1750），巴比伦城邦第一王朝的第六任国王。

[34] Sargon II（约公元前765—公元前705），亚述帝国的国王。

[35] Nebuchadnezzar II（约公元前634—约公元前562），新巴比伦王国君主。

[36] John Edensor Littlewood（1885.06.09—1977.09.06），英国数学家，与哈代合作了100多篇论文。

[37] Aeschylus（约公元前525—约公元前456），古希腊悲剧诗人，有"悲剧之父"的美誉。

[38] Archimedes of Syracuse（约公元前 287—约公元前 212），古希腊数学家、工程学家、天文学家，他被认为是迄今为止最出色的数学家之一。

[39] Og，根据希伯来圣经，是巴山的亚摩利王。

[40] Ananias，圣经中的人物，因欺骗圣灵而死。

[41] Lucius Junius Gallio Annaeanus（约公元前 5—约公元 65），古罗马参议员，作家塞内加（Lucius Annaeus Seneca）的哥哥。

[42] Michel Rolle（1652.04.21—1719.11.08），法国数学家，他发现的罗尔定理是微分学里的一条重要定理。

[43] John Farey（1766—1826.01.06），英国地质学家、作家，1816 年发现"法里数列"。

[44] Charles Haros（17—18 世纪），法国几何学家，"法里数列"的最早发现者。

[45] John Allsebrook Simon（1873—1954），英国政治家，曾担任过数个内阁职务。

[46] William Maxwell Aitken（1879—1964），英国报刊发行人、幕后政治家。

[47] Bertrand Arthur William Russell（1872.05.18—1970.02.02），英国哲学家、数学家和逻辑学家。

[48] 莎士比亚《理查二世》，第三幕第二场。

[49] Alfred North Whitehead（1861.02.15—1947.12.30），英国数学家、哲学家，与罗素合著的《数学原理》是逻辑研究的里程碑。

[50] 怀特海的原文是"错误的文学传统"（an erroneous literary tradition）。

[51] Lancelot Thomas Hogben（1895.12.09—1975.08.22），英国实验动物学家、医学统计学家、科学史家，著有畅销科普书《大众数学》（*Mathematics for the Million*，1936）和《大众科学》（*Science for the Citizen*，1938）。

[52] Henry Ernest Dudeney（1857.04.10—1930.04.23），英国 19 世纪末 20 世纪初最伟大与知名的趣题设计家与娱乐数学家，与同时期的美国趣题奇才萨姆·劳埃德（Sam Loyd）齐名。

[53] Hubert Phillips（1891.12.13—1964.01.09）的笔名，英国经济学家、新闻工作者、出谜人。

[54] Herbert Spencer（1820.04.27—1903.12.08），英国哲学家、社会学家、教育家。

[55] Plato（约公元前 429—公元前 347），古希腊哲学家。

[56] Frederick Soddy（1877.09.02—1956.09.22），英国放射化学家。

[57] William Shakespeare（1564.04.26—1616.04.23），英国剧作家、诗人。

[58] Thomas Otway（1652.03.03—1685.04.14），英国剧作家、诗人。

[59] 莎士比亚《麦克白》，第三幕第二场。

[60] Pierre de Fermat（1601.08.17—1665.01.12），法国著名数学家，被誉为"业余数学家之王"。

[61] Euclid（公元前300年左右），古希腊数学家，《几何原本》的作者。

[62] 即勾股定理。

[63] 参见《数学天书中的证明（第五版）》第4章"表自然数为平方和"，高等教育出版社，2016年3月。

[64] Georg Ferdinand Ludwig Philipp Cantor（1845.03.03—1918.01.06），德国数学家，集合论创始人。

[65] 即素数的个数是无限的。

[66] Theodorus of Cyrene（约公元前465—约公元前399），古希腊数学家。

[67] Eudoxusof Cnidus（约公元前390—约公元前337），古希腊天文学家和数学家。

[68] Walter William Rouse Ball（1850.08.14—1925.04.04），英国数学家。

[69] Harold Scott MacDonald Coxeter（1907.02.09—2003.03.31），英国几何学家。

[70] 这里指大于1的整数。

[71] 中译本（何钦译，商务印书馆1997年7月）第22页："数学的肯定性建筑在它完全抽象的一般性上。"

[72] Diophantus of Alexandria（约公元200—约公元285），古希腊数学家。

[73] Leonhard Euler（1707.04.15—1783.09.18），瑞士数学家、自然科学家。

[74] 指英国科学促进会（British Association for the Advancement of Science），部门A是"数学和物理"学部。

[75] James Clerk Maxwell（1831.06.13—1879.11.05）英国物理学家、数学家。

[76] Arthur Stanley Eddington（1882.12.28—1944.11.22），英国天文学家、物理学家、数学家。

[77] Paul Adrien Maurice Dirac（1902.08.08—1984.10.20），英国理论物理学家，量子力学的奠基者之一。

[78] 克诺索斯是克里特岛上的一座米诺斯文明遗迹。1878 年，英国考古学家阿瑟·伊文思最先对它进行了完整的发掘。

[79] Jules Henri Poincaré（1854.04.29—1912.07.17），法国数学家、天体力学家、数学物理学家、科学哲学家。

[80] 见第 6 节。

[81] John Burdon Sanderson Haldane（1892.11.05—1964.12.01），出生于英国牛津，印度生理学家、生物化学家、群体遗传学家。

[82] Callinicus，希腊语"美丽的胜利者"。

[83] Marie Corelli（1855.05.01—1924.04.24），英国小说家。

[84] Augustus Edward Hough Love（1863.04.17—1940.06.05），英国应用数学家，哈代的导师。

[85] Marie Ennemond Camille Jordan（1838.01.05—1922.01.20），法国数学家。

[86] Lev Nikolayevich Tolstoy（1828.09.09—1920.11.20），俄国文豪。

[87] 特拉法尔加广场，英国伦敦著名广场，坐落在伦敦市中心。

[88] Horatio Nelson（1758.09.29—1805.10.21），英国皇家海军指挥官，地中海舰队总司令。

[89] John Millington Lomas（1917—1945），见本书献辞。

[90] 人民邮电出版社 2017 年 3 月出版。

[91] 人民邮电出版社 2019 年 8 月出版。

[92] Sir Michael Francis Atiyah（1929.04.22—2019.01.11），英国数学家。